PATHOBIOLOGY OF MARINE MAMMAL DISEASES

Volume II

Editor

Edwin B. Howard, D.V.M., Ph.D.

Department of Pathology
University of Southern California
and
Comparative Medical and Veterinary Services Division
Los Angeles County Department of Health Services
Los Angeles, California

CRC Press
Taylor & Francis Group
Boca Raton London New York

CRC Press is an imprint of the
Taylor & Francis Group, an **informa** business

First published 1983 by CRC Press
Taylor & Francis Group
6000 Broken Sound Parkway NW, Suite 300
Boca Raton, FL 33487-2742

Reissued 2018 by CRC Press

Library of Congress Cataloging in Publication Data
Main entry under title:

Pathology of marine mammal diseases.

Bibliography: p.
Includes index.
1. Marine mammals--Diseases. I. Howard, Edwin B.,
1930-
SF997.5.M35P37 1983 599.5'0421 82-20611
ISBN 0-8493-6311-8 (v. 1)
ISBN 0-8493-6311-6 (v. 2)

A Library of Congress record exists under LC control number: 82020611

Publisher's Note
The publisher has gone to great lengths to ensure the quality of this reprint but points out that some imperfections in the original copies may be apparent.

Disclaimer
The publisher has made every effort to trace copyright holders and welcomes correspondence from those they have been unable to contact.

ISBN 13: 978-1-315-89631-1 (hbk)
ISBN 13: 978-1-351-07541-1 (ebk)

Visit the Taylor & Francis Web site at http://www.taylorandfrancis.com and the
CRC Press Web site at http://www.crcpress.com

THE EDITOR

Edwin B. Howard, D.V.M., Ph.D. is Associate Professor of Pathology, University of Southern California School of Medicine, and serves as Chief of the Los Angeles County/ U.S.C. Section of Comparative Pathology and as Chief of Laboratory and Disease Investigation Division, Comparative Medical and Veterinary Services Division of the Los Angeles County Department of Health Services.

Dr. Howard received his B.S. and D.V.M. degrees from Colorado State University in 1957 and 1959 respectively. After an internship he continued postgraduate studies in pathology and virology at Colorado State University and then at the University of Connecticut, where he received a Ph.D. degree in pathology in 1966. He became a Diplomate of the American College of Veterinary Pathologists in 1965.

Professional positions prior to affiliating with the University of Southern California and Los Angeles County Department of Health Services in 1974, included appointments with Battelle-Northwest Laboratories, the State of Hawaii and University of Hawaii School of Medicine, the Veterans Administration, and the New Jersey College of Medicine.

Dr. Howard has served as consultant and member of the Late Effects of Radiation Advisory Committee to the U.S. Public Health Service, and member of a similar committee with the Environmental Protection Agency. He has served for several years as Co-Director of the Marine Biomedical Research Center at the University of Southern California, and as member of the Advisory Board of the Charles L. Davis, D.V.M. Foundation.

Professional affiliations include the American College of Veterinary Pathologists, American Veterinary Medical Association, American Association of Pathologists, International Academy of Pathology, Society of Toxicologic Pathologists, Radiation Research Society, American Association of Veterinary Laboratory Diagnosticians, and the Wildlife Disease Association.

CONTRIBUTORS

James O. Britt, Jr., D.V.M.
Veterinary Pathologist
Comparative Medical and Veterinary Services Division
Los Angeles County Department of Health Services
Downey, California

Lanny Cornell, D.V.M.
Senior Vice-President
Sea World Enterprises
San Diego, California

Edwin B. Howard, D.V.M., Ph.D.
Associate Professor of Pathology
Department of Pathology
University of Southern California School of Medicine
Los Angeles, California
and
Comparative Medical and Veterinary Services Division
Los Angeles County Department of Health Services
Downey, California

Rene Itahara
Laboratory Technologist
Comparative Medical and Veterinary Services Division
Los Angeles County Department of Health Services
Downey, California

Sidney R. Jones, Colonel, U.S.A.F., B.S.C. (V.C.)
Department of Veterinary Pathology
Armed Forces Institute of Pathology
Washington, D.C.

George K. Matsumoto, B.S., M.T.
Public Health Microbiologist
Comparative Medical and Veterinary Services Division
Los Angeles County Department of Health Services
Downey, California

George Migaki, D.V.M.
Chief Pathologist
Registry of Comparative Pathology
Armed Forces Institute of Pathology
and
Universities Associated for Research and Education in Pathology, Inc.
Washington, D.C.

Clara Nagano, B.S., M.T.
Microbiologist
Veterinary Laboratory Branch
State of Hawaii
Department of Agriculture
Honolulu, Hawaii

John G. Simpson, D.V.M.
Department of Pathology
University of Southern California School of Medicine
Los Angeles, California
and
Animal Pathology Service
Camarillo, California

TABLE OF CONTENTS

Volume I

TABLE OF CONTENTS

Volume II

Chapter 1

MYCOTIC DISEASES IN MARINE MAMMALS

George Migaki and Sidney R. Jones

TABLE OF CONTENTS

I. INTRODUCTION

During the past two decades there has been increased use of marine mammals for education, and the presence of these animals is now commonplace in many aquariums, oceanariums, zoos, and aquatic animal research institutions.[1,2] Also during this period, increased medical knowledge concerning the diseases and therapy of these aquatic animals has been gained.[3,4] Although reports of mycotic infections in marine mammals are relatively rare compared to those of infectious diseases caused by viruses, mycoplasmas, bacteria, and helminthic parasites, it is important to remember that fungal organisms can contribute to the death of these animals.[5,6]

Except for dermatophytosis or ringworm, lesions produced by fungi often are severe and may be systemic involving various organ systems; therapy, in most instances, has not been successful. Also, mycotic infections generally are not considered contagious and there is little or no evidence of the infection being spread from one animal to another, except for some of the dermatophytoses. It is believed that infection in animals generally occurs by accidental contact with fungi in the soil or in organic debris where they exist and grow as saprophytes.[7-9] Infection may occur by inhalation of the fungal spores resulting in a primary lesion in the lung. Sometimes infection follows implantation of the fungal spores on abraded areas of the skin or mucous membrane. It should be kept in mind that many, though not all, fungi are secondary or opportunistic organisms; therefore, predisposing factors play a major role in the development of mycotic infections.[10] Among the predisposing factors or conditions associated with suppression of the immune system are stress to the animals from being placed in captivity; deficiencies resulting from malnutrition; the prolonged use of broad-spectrum antibiotics or corticosteroids; congenital defects of the immune system; pre-existing diseases such as anemias or malignant neoplasms; and coexisting septicemias caused by toxic, physical, or infectious agents.[11]

Much of our current knowledge on mycotic infections in marine mammals stems from our experience with such diseases in other animals[9,11-18] and humans[10,19-22] and we encourage readers to consult the cited references.

The following procedures for the identification of a fungus as the etiologic agent are recommended:[23,24]

1. Inoculation of the suspected fungus in laboratory animals in an attempt to reproduce the disease
2. Preparation of smears for microscopic examination after treating the fresh specimen with potassium hydroxide and lactophenol cotton blue and with Gram's and acid-fast stains
3. Isolation and identification of the fungal organisms on artificial media with and without antibacterial and antifungal compounds
4. Fluorescent antibody techniques on fresh or formalin-fixed specimens when specimens for cultural studies are unavailable[25]
5. Preparation of tissue sections stained with hematoxylin and eosin and other special histochemical procedures such as periodic acid-Schiff (PAS), Gomori methenamine-silver (GMS), Gridley, Mayer's mucicarmine, Fite-Faraco acid-fast, and Brown-Brenn Gram.[26]

From a diagnostic standpoint, mycotic infections can be divided into two groups: deep (or systemic), which are severe pyogranulomatous lesions, and superficial, those that are confined to the skin and cause relatively mild lesions. Although nocardiosis, actinomycosis, and dermatophilosis are caused by actinomycetes and not fungi, they traditionally are grouped under mycotic infections since the lesions they produce are similar to those caused by fungi.

Botryomycosis, caused by a nonfilamentous bacterium, also is included because the lesions are similar to those caused by fungi and actinomycetes. This report does not include the mycotoxicoses, which are diseases resulting from the ingestion of contaminated feeds containing toxins produced by fungi.

II. CANDIDIASIS

Candidiasis, known also as moniliasis and thrush, is caused by a yeast belonging to the genus *Candida*, the most common being *Candida albicans*. Other species include *C. glabrata, C. krusei, C. tropicalis*, and *C. parapsilosis. C. albicans* is worldwide in distribution and can be found in small numbers on the mucous membrane of the alimentary and genital tracts and on the skin of normal animals without any evidence of clinical disease. *C. albicans* differs from some of the other fungi in that it does not appear to grow in inanimate matter. However, when the immune system of an animal is suppressed, *Candida*, like some of the other opportunistic fungi, may proliferate and cause severe local infection in the mucous membrane and skin. The cutaneous and mucosal lesions appear as whitish, creamy plaques; in systemic infections the affected organs have discrete whitish areas due to necrosis and leukocytic infiltrates. Histologically, large colonies of septate hyphae, pseudohyphae (3 to 4 μm in diameter) and blastospores (3 to 6 μm in diameter) are important morphologic criteria for the recognition of *Candida* (Figure 1). Cultural studies are necessary for species identification.

In lesions involving the skin and mucous membrane, *Candida* is found in the epithelial tissue which may have undergone acanthosis with pseudoepitheliomatous hyperplasia. In systemic candidiasis, any organ may be involved, but the kidneys, heart, and lungs commonly are affected. Blastospores appear to be more numerous than hyphal structures in organs affected by systemic infections.

In animals, the alimentary tract, especially the oral mucosa and esophagus and the crop in birds, is a common site of lesions. Candidiasis appears to be more common than some of the other mycotic infections in marine mammals. It has been a problem in marine mammals in seaquariums. The clinical disease may be associated with prolonged use of antibiotics and corticosteroids or overtreatment of the water for purity, or both.[6] Among reports of candidiasis in pinnipeds is one involving an adult male harbor seal (*Phoca vitulina*) kept in fresh water in a zoological park in Japan. The seal had multiple large areas of alopecia and erythema of the eyelids, lips, limbs, and abdomen.[27] Three elephant seals (*Mirounga leonina*) in a zoo in Stuttgart, Germany, had multiple cutaneous lesions of candidiasis on the caudal fin, eyelids, lips, and perianal areas.[28] Reports of *Candida* infections involving cetaceans are numerous, including one report of five freshwater dolphins (*Inia geoffrensis*) at the Duisburg Zoological Garden, Germany.[29] Several weeks following capture near Venezuela, a dark grey film covered the entire body surface of these dolphins. In addition to *C. albicans*, colonies of bacteria, including the genera *Streptococcus, Staphylococcus, Pseudomonas, Escherichia, Aerobacter*, and *Proteus* were isolated from the grey film. Three performing bottlenosed dolphins (*Tursiops truncatus*) in Buffalo, New York, each about 10-years-old, had severe ulcerative dermatitis that involved much of their bodies.[30] Two of the dolphins also had extensive esophagogastric ulcers due to *Candida*, which was believed to be the cause of their deaths. The pathogenesis of the cutaneous and esophageal lesions was undetermined; however, it was suggested that environmental factors such as captivity, performing, water quality, and persistent unknown infection could have suppressed the immune system. Candidiasis was diagnosed in the terminal esophagus (forestomach) of three captive Atlantic bottlenosed dolphins (*Tursiops truncatus*) by the use of a flexible fiberoptic endoscope.[31] Inflammation without ulcers was noted in one animal; healed ulcers were found in two animals. In the Registry's files are systemic lesions of candidiasis in two killer whales

(*Orcinus orca*). Specimens were contributed by Dr. Lanny Cornell. Visceral lesions were found in many organs including the heart, kidneys (Figure 2), and lymph nodes. In a survey of the causes of death of seventeen killer whales in six major oceanariums in North America, candidiasis was reported to be the cause of death in a young male killer whale that had anemia.[32] Additional cases of candidiasis in cetaceans and pinnipeds are listed by Sweeney et al.[5] in their review of systemic mycoses.

C. albicans was isolated from the tongue and stomach of a captive two-week-old polar bear (*Thalarctos maritimus*) that died at the Calgary Zoological Society.[33] The dorsum of the tongue was hyperkeratotic and parakeratotic, and a large infarct was found in the stomach.

III. SPOROTRICHOSIS

Sporotrichosis is caused by *Sporothrix (Sporotrichum) schenckii*, a ubiquitous saprophyte in nature which can be isolated from plants, trees, moss, and soil. The fungus is dimorphic. Infection generally follows a puncture wound or abrasion of the skin allowing implantation of the fungal spores in the subcutaneous tissue. Dissemination of the infection occurs along the lymph vessels resulting in subcutaneous nodules along the lymph vessels with metastatic lesions developing in the regional lymph nodes and other organs. Primary pulmonary infection may occur following inhalation of the fungal spores.

Histologically, a pyogranulomatous inflammatory process characterized by large numbers of histiocytes and small foci of polymorphonuclear neutrophils is a consistent finding. *S. schenckii* can be found free or phagocytosed by histiocytes. They vary from round to ovoid to cigar-shaped yeast cells, and they measure 4 to 10 μm in diameter. Budding forms occur, resulting in short chains of cells resembling pseudohyphae. The morphology of *S. schenckii* in tissue sections is best demonstrated when stained with PAS or GMS.

Sporotrichosis has a wide geographic distribution and has been reported in many countries and in many animal species[12,16] and humans.[10] In animals, at least during the past 20 years, most reports have been in horses, dogs, cats, and pigs.[34] A single report exists of the disease in a marine mammal, a Pacific white-sided dolphin (*Lagenorhynchus obliquidens*) of undetermined age that had been in captivity for about three years at Sea World, Orlando, Florida.[35] During this period the animal had erratic eating habits and had been treated symptomatically with various types of antibiotics. Lesions at necropsy were cutaneous lacerations over the body, enlarged and hemorrhagic respiratory lymph nodes, and hemorrhagic lesions in the lungs and oral mucous membrane. Histologically, a severe serofibrinous necrotic granulomatous lymphadenitis and unusually large numbers of *S. schenckii* were present which suggested that the immune system of the animal was suppressed (Figure 3). The origin of the infection was undetermined; however, the skin was suspected in this animal despite the fact that skin lesions were unavailable for histologic or cultural examination.

IV. COCCIDIOIDOMYCOSIS

Coccidioidomycosis, known also as coccidioidal granuloma, San Joaquin Valley disease or desert fever, is caused by *Coccidioides immitis*, a dimorphic fungus. This is a dust-borne infection, and inhalation of the fungal spores is the only proven method of infection. *C. immitis* is a soil-inhabitant and grows as a saprophyte in the semi-arid areas of North, Central, and South America. In the U.S. this area, called the Lower Sonoran Life Zone, encompasses the desert regions of Texas, New Mexico, Arizona, California, Nevada, and Utah.[36] Infective areas continue southward into the deserts of Mexico, Central America, Venezuela, and the Gran Chaco-Pampa region of Paraguay, northern Argentina and Bolivia. The hot, dry summers, mild winters, scanty rainfall, alkaline soil, and sparse vegetation are conducive to the growth of *C. immitis*. From an epizootiologic standpoint, the disease has a limited geographic distribution when compared to other mycotic infections.

Histologically, a common finding is a suppurative granulomatous inflammatory process characterized by foci of polymorphonuclear neutrophils surrounded by a wide zone of histiocytes, multinucleated giant cells, lymphocytes and fibrous tissue. *C. immitis* is a relatively large round yeast cell varying from 20 to 80 μm in diameter. The cell wall is thick and is "doubly contoured". Mature spherules contain endospores that vary from 2 to 5 μm in diameter. Rupture of the mature spherules attracts neutrophils.

C. immitis has a broad host range, and the disease has been reported in many animal species.[12,16,36] The disease is primarily a respiratory disorder; however, since animal species vary in their susceptibility to *C. immitis,* clinical signs are dependent upon the severity of the pulmonary lesions and dissemination to other organ systems. For example, in cattle the lesions are confined to the lungs and lymph nodes, whereas in dogs and horses the lesions are disseminated throughout the body. There are two reports of coccidioidomycosis in marine mammals. The first involves an adult male California sea lion (*Zalophus californianus*) that had been housed in a zoo in Tucson, Arizona for approximately five years.[37] No clinical signs were noted except that it had been anorexic for several days prior to death. Multiple suppurative granulomatous lesions were found throughout the liver and spleen (Figure 4). Since other organs were not available for examination, the extent and distribution of the lesions could not be determined. The second involves a weak and emaciated adult male sea otter (*Enhydra lutris*) that was found stranded on Atascadero Beach in Morro Bay, California.[38] Despite three weeks of therapy in captivity the animal died. Disseminated lesions were found in the lungs, lymph nodes, liver, and spleen.

V. NORTH AMERICAN BLASTOMYCOSIS

North American blastomycosis, known also as Gilchrist's disease, is caused by *Blastomyces dermatitidis*, a dimorphic fungus. Infection occurs as a result of inhalation of spores of *B. dermatitidis*, which proliferate as saprophytes in nature; however, their natural habitat remains unknown. A primary lesion develops in the lungs with subsequent metastases to various organs including the skin. Clinical signs in animals are variable and are dependent on the severity of the lesions and the organs affected. Histologically, a pyogranulomatous inflammatory process is seen in which *B. dermatitidis* occurs either free or phagocytosed in macrophages as large round yeast cells, 8 to 25 μm in diameter, with a single broad-based bud. The cell wall is relatively thick and has the appearance of being "doubly contoured." These morphologic characteristics are best seen when tissue sections are stained with GMS.

The disease has been reported in various animals including the dog, horse, and cat in North and Central America and Africa.[12,16] In the U.S. the disease is seen mostly in dogs residing in states bordering the Ohio, Mississippi, and Missouri river valleys.[39] Among reports of the disease in marine mammals is that of an eight-year-old female Steller's sea lion (*Eumetopias jubata*) with disseminated lesions in lungs, lymph nodes, spleen, liver, thyroid, and kidneys.[40] The animal, housed at the Chicago Zoological Park for six years, had been reported to be in good health for most of that period; however, several weeks before death, she spent more time than usual resting out of the water. Several days prior to death, she became very weak and developed anorexia. In a review of systemic mycoses, Sweeney et al.[5] list blastomycosis in a subadult bottle-nosed dolphin (*Tursiops truncatus*) that had been captured on the Mississippi Gulf Coast and maintained in captivity for six months in Hawaii. No mention was made of the anatomic location or severity of the lesions.

VI. CRYPTOCOCCOSIS

Cryptococcosis is caused by an encapsulated yeast, *Cryptococcus neoformans*, which occurs as a saprophyte in nature. Although the source of infection has not been identified, a high correlation has been found between *C. neoformans* and nests and droppings of pigeons, thus suggesting that avian excrements provide a suitable substrate for growth.[41] In the past it was believed that pigeons resist infection to *C. neoformans* but recently it has been shown that these birds are susceptible to such infection.[42]

The lungs are generally the primary site of infection following inhalation of *C. neoformans*. Dissemination to other organs, especially to the central nervous system, may occur. An exception is in cattle. Several reports exist[43,44] of primary mastitis due to *C. neoformans* in dairy herds in which the infection resulted from use of contaminated milking machines or contaminated intramammary medicinal preparations. On gross examination, the lesions often have a slimy consistency.

Histologically, moderate histiocytic response occurs, though in some anergic hosts little leukocytic infiltrate is present. *C. neoformans* is round, 5 to 10 μm in diameter, with a single bud. A wide clear area surrounding *C. neoformans* represents a mucinous capsule which can be demonstrated best with Mayer's mucicarmine stain.

Cryptococcosis has a worldwide distribution, and the disease has been reported in many animal species.[12,16] There is a report in marine mammals involving a seven-year-old male Atlantic bottle-nosed dolphin (*Tursiops truncatus*) that had been held in captivity for about four years in California.[45] About a month before its death, moderate rales in both hemithoraces and loss of weight were noted. At necropsy a solitary firm subpleural nodule due to *C. neoformans* was found in the left lung. Death was attributed to septicemia resulting from a perforated ulcer in the pyloric region of the stomach. Although the source of *C. neoformans* was not determined, many seafaring birds were observed roosting near the compound where the dolphin was kept in captivity. Accessioned in the Registry's files are specimens from a female Pacific white-sided dolphin (*Lagenorhynchus obliquidens*) that had been in captivity at Sea World, Aurora, Ohio, She had been undergoing antibiotic therapy for elevated white blood cell count prior to death. Histologically, myriads of *C. neoformans* were observed in the lungs (Figure 5) and lymph nodes with little or no leukocytic response, indicating a profuse multiplication of *C. neoformans*. Tissues and clinical history were provided by Dr. Robert O. Stevens.

VII. HISTOPLASMOSIS

Histoplasmosis, known also as Darling's disease, is probably the most common of the systemic mycoses and is caused by the yeast, *Histoplasma capsulatum*. This organism grows in soil enriched by fecal droppings of chickens, other birds, and bats.[46] When this soil is turned, fungal spores are scattered in the atmosphere. The usual route of infection is by inhalation with the primary lesions in the lungs from which dissemination may occur to other parts of the body.

Since the fungus has a predilection for the reticuloendothelial system, the outstanding histologic finding is the proliferation of histiocytes. These cells are enlarged and their cytoplasm contains large numbers of *H. capsulatum*. The histiocytic proliferation in some cases may be so extensive that the lesions resemble a lymphoreticular neoplasm. *H. capsulatum* organisms are round to ovoid and 2 to 4 μm in diameter. They have a centrally placed nucleus and a thick cell wall but no capsule. These morphologic features are best demonstrated when stained with GMS.

Histoplasmosis has a worldwide distribution. In the U.S., prevalence of the disease appears to be highest in the Central Mississippi River Valley, the Ohio River Valley, and the areas along the Appalachian Mountains.[8] The disease has been reported in a wide variety of animals.[12,16,47] However, the high body temperature of the chicken and other avian species apparently prevents them from becoming infected. In marine mammals, histoplasmosis has been reported only in an adult female harp seal (*Pagophilus groenlandicus*) that had been in captivity for seven years.[48] The animal had been used in acoustic experiments but two to three months prior to death, she had nasal discharge and experienced respiratory distress. Two large submandibular masses and multiple nodules in the lungs were found (Figure 6). The source of infection was not determined in this seal; however, sea gulls on nutritional studies were housed within concrete tanks in wire-enclosed pens approximately 75 feet from where the seal was maintained. Whether the sea gulls contributed to the source of infection in the harp seal remains unknown.

VIII. ASPERGILLOSIS

Aspergillosis is caused most commonly by *Aspergillus fumigatus*, less commonly by other species, such as *A. niger* and *A. terreus*. It occurs worldwide and generally affects the respiratory tract. The placenta, gastrointestinal tract, and skin also are sites of infection. The fungi are abundant in the environment where they live as saprophytes in the soil and in dead plant and animal matter. Infection generally occurs by inhalation or by direct contact with the fungi. Large areas of necrosis and a granulomatous inflammatory response are the principal pathologic findings. Histologically, the fungi usually are present in large numbers and are identified as septate hyphae of uniform diameter (about 3 to 6 μm) with dichotomous branching, usually at about a 45° angle. In chronic lesions, an amorphous eosinophilic precipitate (Splendore-Hoeppli phenomenon)[49] may be found surrounding the fungi.

Avian species most commonly are affected and the respiratory tract is the principal site of infection. Abortions have been reported in cattle, sheep, and horses. The Registry's files contain two cases of aspergillosis in marine mammals, both Atlantic bottle-nosed dolphins (*Tursiops truncatus*). One animal had large discrete granulomatous lesions in the lungs and a large nodule protruding on the mucosal surface of the trachea (Figure 7). This case was contributed by Dr. Edwin Howard, Downey, California. The lesions were of long duration as evidenced by fibrosis and eosinophilic precipitate surrounding the fungi. The other animal was from Sea World, Orlando, Florida and had a granulomatous pneumonia. The lesions appeared to be of shorter duration and no eosinophilic precipitates were found around the fungi. The specimens were contributed by Dr. D. O. Buesse. Pulmonary aspergillosis with extensive distribution of lesions throughout the lungs has been reported in an adult female Atlantic bottle-nosed dolphin (*T. truncatus*) from Florida.[50] She had been captured at a young age (about two-years-old) and had been a performer for 13 years. Three days prior to death she developed labored breathing. Sweeney et al.[5] in their review on systemic mycoses have listed several other cetaceans and pinnipeds affected with aspergillosis.

IX. ZYGOMYCOSIS

Zygomycosis is the term now used to describe the diseases formerly called by such names as mucormycosis, phycomycosis, entomophthoromycosis, hyphomycosis, basidiobolomycosis, oomycosis, subcutaneous phycomycosis, rhinophycomycosis, phycomycosis entomophthorae, and rhinoentomophthoromycosis.[17] It includes a diverse group of mycoses caused by a variety of fungi including the genera *Absidia, Rhizopus, Mucor, Basidiobolus, Conidiobolus (Entomophthora), Hyphomyces, Mortierella, Cunninghamella*, and *Saksenaea*.[10,17] The organisms are ubiquitous soil inhabitants. These agents are opportunistic organisms, and infection may result from predisposing factors such as traumatic damage to the tissue or suppression of the immune system from prolonged use of corticosteroids, antibiotics or irradiation therapy or some other pre-existing disease. Clinical manifestations vary depending upon the organs and tissue structures involved and the severity of the infection. The lesions in some of the diseases are confined to the skin and subcutaneous tissue while in others, systemic infection occurs.

Some of the fungi such as *Mucor* have a propensity for invading blood vessels resulting in thrombosis, ulceration, and infarction. In some infections an extensive caseous granulomatous inflammatory process develops. In tissue sections, the fungi appear as broad, usually nonseptate, branching hyphae with bulbous enlargements. In infections caused by the genus *Conidiobolus (Entomophthora)*, a large amount of bright eosinophilic amorphous precipitate (Splendore-Hoeppli phenomenon)[49] surrounds the hyphal structures.

Zygomycosis has been reported in many species with the infection involving the gastrointestinal tract, skin, and nasal cavity.[12,16,17] Relatively few reports exist of the disease in marine mammals. A female harp seal (*Pagophilus groenlandica*) captured as an infant near Nova Scotia, Canada and held in a freshwater pool at the Bronx Zoological Park for about two years became anorexic, listless, and died.[51] The significant pathologic finding was a granulomatous pneumonia caused by *Mucor pusillus*. Reported cases of the disease in cetaceans include a young male southern right whale (*Eubalaena australis*), about three to four-months-old, beached along the coast of South Africa.[52] Several large suppurative granulomatous lesions containing non-septate hyphal organisms morphologically similar to the genus *Mucor* were found in the left epiaxial muscle. In Florida, an adult Atlantic bottle-nosed dolphin (*T. truncatus*) housed in an oceanarium had clinical signs of vomiting and respiratory difficulties several months prior to death. Numerous large granulomatous nodules containing organisms morphologically resembling the genus *Entomophthora* and surrounded by amorphous eosinophilic precipitate were found in the lungs.[53]

Accessioned in the Registry's file is a case of systemic zygomycosis in an 18-year-old female killer whale (*Orcinus orca*). A resident of the Vancouver Public Aquarium for 13 years, she became anorexic and less active about three weeks prior to death. Pathologic findings included mycotic invasion of the blood vessels of many organs, including the gastrointestinal tract, lungs, heart, and skeletal muscle, resulting in thrombosis, embolism, and infarction (Figure 8). Specimens were contributed by Dr. A. C. MacNeill.

X. LOBOMYCOSIS

Lobomycosis, also known as Lobo's disease or keloidal blastomycosis, is caused by *Loboa loboi*. Since the lesions are restricted to the skin with no evidence of dissemination to the regional lymph nodes or other organs, it is believed that infection follows abrasion or traumatization of the skin.[10,54] The gross lesions are multiple, elevated, whitish, crusty, nodular growths on the skin. Due to the multiple locations of the lesions on the body, it is believed that traumatic autoinoculation takes place.

Histologically, the lesions, found only in the dermis of the skin, cause a marked acanthosis characterized by proliferation of the squamous epithelial cells and irregular downgrowths of the rete pegs. The granulomas in the dermis are composed almost entirely of histiocytes, many of which have coalesced forming multinucleated giant cells. Fungal structures of *L. loboi* are numerous and are found both free and phagocytosed by histiocytes and giant cells. They are round to lemon-shaped, 5 to 10 μm in diameter, connected by short tubes, and arranged in long branching chains resembling "tinker toys". The morphologic features can be demonstrated best with GMS or Gridley fungus stains (Figure 9). In humans, lobomycosis is considered a geographic disease as it has been reported only in natives living in tropical forests of south and central America. A high frequency of the disease is reported in certain parts of Brazil, Surinam, and French Guiana.[19,54] Efforts to reproduce the disease by experimental inoculation of animals with isolated fungal cells have been unsuccessful. Dolphins are the only animals in which lobomycosis has been reported.[55-60] The gross and histologic lesions are identical to those in humans, including confinement to the skin. All of the dolphins were identified as the Atlantic bottle-nosed dolphin (*Tursiops truncatus*) except for one[57] which was a Guiana dolphin (*Sotalia guianensis*). Except for the Guiana dolphin which was captured in the estuary of the Surinam River, all of the other dolphins were captured along the coast of Florida. The geographic locations include Sarasota,[55] Marineland,[56] San Marco's Island,[58] St. Petersburg,[59] and Vero Beach.[60] The anatomic location of the lesions varied in these dolphins but they were found mainly on the head, flippers, abdomen, fins, back, tail stocks, and flukes.

How dolphins contract the disease, which previously had been found only in humans, is unknown. The involvement of only dolphins and humans in this geographic area suggests that lobomycosis is an enzootic disease, but the question as to why other animal species have not been infected spontaneously continues to be a mystery.

XI. FUSARIOMYCOSIS

Fusariomycosis is caused by the genus *Fusarium* which is a common soil inhabitant and considered to be saprophytic. Reports of this disease in animals and humans are relatively rare. *Fusarium* has been incriminated as the cause of keratitis in humans.[20] Infections of the skin and eyes of reptiles have been reported.[61] The only report of the disease in marine mammals involves three California sea lions (*Zalophus californianus*) and three gray seals (*Halichoerus grypus*) housed at the National Zoological Park.[62] They were affected with severe proliferative dermatitis. The lesions were firm raised papules found mainly on the face, trunk and flippers. Histologically, acanthosis, hyperkeratosis, and intraepithelial abscesses were the principal findings (Figure 10). *Fusarium* in tissue sections appeared as septate branching hyphae varying from 2 to 5 μm in diameter. The morphologic features of the fungi were best demonstrated when stained with PAS. In the gray seals, lesions appeared more severe in the hot summer months and showed regression during the winter months. Predisposing factors responsible for infection remain undetermined but the onset of the disease may be associated with some factor related to captivity such as initial excessive chlorination or high temperature of the water in the pool.

XII. DERMATOPHYTOSIS

Dermatophytosis, known also as ringworm, is caused by fungi that parasitize the keratinized layer of the epidermis, hair, feathers, and horns of animals. The deeper tissues, i.e., dermis and subcutaneous tissue, of the skin generally remain unaffected. Three genera are recognized, *Microsporum*, *Trichophyton*, and *Epidermophyton*. Dermatophytosis occurs worldwide and has been reported in many animal species. Because of their widespread distribution in both host specificity and natural habitat, the fungi are divided into three groups: (1) anthropophilic, those that cause ringworm in humans with infection occurring by direct contact; (2) zoophilic, those that have adapted themselves to specific animal species with infection taking place by direct contact or contaminated fomites, and; (3) geophilic, those that normally inhabit the soil with infection taking place by direct contact.

Reports of dermatophytosis are relatively rare in marine mammals. In Japan numerous discrete conical nodules were found on the posterior part of the trunk of a captive Atlantic bottle-nosed dolphin (*Tursiops truncatus*).[63] The fungus was identified as belonging to the genus *Trichophyton*. A captive manatee (*Trichechus* sp.) from Florida had numerous erosions on the skin of its nose, face, flippers, and tail.[64] The fungus was identified as *Epidermophyton floccosum*. A report of the disease in pinnipeds involves a 9-year-old female harbor seal (*Phoca vitulina*) that had been housed in a zoo in St. Paul, Minnesota for five years. She had generalized thinning of the hair and alopecia about the face and nose; the lesions later developed into a pustular dermatitis.[65] *Microsporum canis* was cultured from the lesions.

XIII. NOCARDIOSIS

Nocardiosis is caused by a bacterium belonging to the aerobic actinomycete, *Nocardia*. Species most frequently incriminated are *Nocardia asteroides, N. brasiliensis*, and *N. caviae*. The organisms are saprophytes in nature. Infection occurs by contact on abraded skin, by inhalation or by ingestion of contaminated feed. The respiratory tract probably is the most common primary site of infection with dissemination by the blood and lymph vessels to other organs such as the brain. Secondary involvement of the nervous system may result in clinical signs of nervous disorders. A suppurative inflammatory process is the usual histologic finding with nocardial organisms diffusely distributed throughout the lesions. In some infections the filamentous organisms proliferate and become organized into granules or grains that are surrounded by granulomatous inflammatory tissue. These findings are called mycetomas. In some instances the granules are surrounded by an eosinophilic amorphous precipitate.

Nocardiosis has been reported in many animal species. A primary suppurative inflammation of the lungs is the usual pathologic finding. The organism in tissue sections is slender (about 1 μm in diameter), filamentous, branching, Gram-positive, and partially acid-fast.

Nocardiosis in marine mammals appears to be relatively common.[5] At a sea life park in Hawaii, three animals were affected.[66] They were a young female pilot whale (*Globicephala scammoni*) that died about two months following capture, a mature male Pacific bottle-nosed dolphin (*Tursiops gilli*) that died eight months following capture, and a mature female Pacific bottlenosed dolphin (*Tursiops gilli*) that died four years following capture. All of the animals had a severe suppurative granulomatous pneumonia. *N. asteroides* was cultured from the pilot whale, *N. brasiliensis* was cultured from one of the dolphins, and *N. caviae* from the other dolphin. The origin of the infections was undetermined but soil contamination of the tank water with *Nocardia* was considered a possibility. Sweeney et al.[5] list three cetaceans from which *N. asteroides* was isolated: (1) an adult male killer whale (*Orcinus orca*) that was captured in Puget Sound and died several years later in captivity in Hawaii; (2) an adult male false killer whale (*Pseudorca crassidens*), and; (3) an adult spinner dolphin (*Stenella longirostris*). The latter two died from beaching in Hawaii. A four-month-old male Beluga whale (*Delphinapterus leucas*) that was born in an aquarium in Vancouver developed nocardiosis.[67] Two to four weeks prior to death, the animal was noted to have aberrant swimming patterns and became progressively weaker. In addition to severe sloughing of epidermis over the anterior half of the body, multiple abscesses occurred in both the thoracic and abdominal viscera (Figure 11). *Nocardia* was cultured. The umbilicus was considered the primary site of infection because of adhesions of the intestine to the abdominal wall in this area. A captive Atlantic bottle-nosed dolphin (*Tursiops truncatus*) from Florida was reported to have ulcerative lesions of the mouth and snout.[68] In the necrotic areas, the organisms were organized into aggregates forming mycetomas. The organisms were identified as *Nocardia paraguayensis*. Also reported from Florida was a beached pygmy sperm whale (*Kogia breviceps*) from which an organism belonging to the "*Actinomyces-Nocardia* group" was found in tissue sections prepared from a lesion in the lung.[66] Accessioned in the Registry's files are specimens from a female Risso's dolphin (*Grampus griseus*) that died from a severe suppurative pneumonia and septicemia following rescue five months earlier along the coast of Florida. The clinical history and specimens were contributed by Dr. Ed Asper, Orlando, Florida.

Several reports of nocardiosis involve pinnipeds. *N. asteroides* was cultured from a young, captive, female harbor porpoise (*Phocaena phocaena*) that was affected with pneumonia, pleuritis, and abscesses of the lymph nodes and kidneys.[66] A report from New Zealand concerns a captive mature male leopard seal (*Hydrurga leptonyx*) that had large multiple suppurative lesions in the subcutaneous tissue, blubber, skeletal muscles, kidneys, and lungs.[69]

XIV. DERMATOPHILOSIS

Dermatophilosis, also known as streptothricosis or mycotic dermatitis, is caused by an aerobic and facultatively anaerobic actinomycete, *Dermatophilus congolensis*. The lesions are self-limiting to the skin with no evidence of metastasis to regional lymph nodes, and the inflammatory process generally does not extend beyond the dermis. Dermatophilosis can be transmitted to other animals, therefore an infected animal serves as a source of infection to others. Infection follows direct contact, and biting insects also have been incriminated in the spread of the disease. Predisposing factors include constant wetness of the skin and cutaneous abrasions. The gross lesion is an exudative dermatitis characterized by multiple raised crusted areas containing matted hair and serofibrinous exudate. In advanced lesions, papillomatous masses composed of keratinaceous material arise. Histologically, the skin is greatly thickened and characterized by hyperkeratosis, parakeratosis, and suppurative exudate.[70] A diagnostic finding is the presence of alternate layers of parakeratotic epithelial tissue and suppurative exudate. This lamination develops because *D. congolensis* invades and multiplies in living and noncornified epithelial cells and its growth attracts polymorphonuclear neutrophils, resulting in suppurative exudate. *D. congolensis* is a Gram-positive, branching filamentous organism with vertical and horizontal septations resulting in coccus-like bodies.

The disease is found worldwide and has been reported in many species[12,16] including coldblooded animals.[71] The two reports of the disease in marine mammals both involve polar bears (*Thalarctos maritimus*).[72,73] Six polar bears housed at the Auckland Zoological Park in New Zealand had a chronic extensive dermatitis.[72] Cutaneous lesions were first seen on the head, thorax, and lumbar regions, later spreading down the sides of the body and on to the legs. Biting flies (*Stomoxys calcitrans*) were numerous and were considered responsible for the perpetuation and spread of the disease from animal to animal. The second report involves two polar bears housed at the Detroit Zoological Park.[73] One was a five-year-old female that had suffered from progressive dermatitis for three years. A crusty, dark gray scab was found over most of the body. The other animal was a 15-month-old female that had yellow staining and an accumulation of black greasy material on the coat which first were noticed about six months earlier.

A single report exists of dermatophilosis in pinnipeds.[74] Two captive South American sea lions (*Otaria bryonia* Blainville) that had been imported to Germany from South America had multiple discrete, elevated, crusty cutaneous lesions throughout the body surface (Figure 12).

XV. ACTINOMYCOSIS

Actinomycosis in animals is caused by an actinomycete, usually *Actinomyces bovis*. Other species such as *A. israeli* and *A. viscosus* less commonly are the cause of the disease. It is important to note that these organisms are commensals in the oral cavity and, being opportunists, will invade previously damaged tissues.

The disease is sporadic and worldwide in distribution. It is a chronic, suppurative disease characterized by abscesses and fistulous tracts draining areas of suppuration. The suppurative exudate generally contains yellowish to yellowish-white firm granules ("sulfur granules") composed of aggregates of gram-positive, branching filamentous organisms, about 1 μm in diameter, which are surrounded by brightly eosinophilic material arranged in radiating clubs (Splendore-Hoeppli phenomenon).[49]

Actinomycosis most commonly involves the osseous structures of the skull of cattle; however, in other domesticated mammals soft tissues are affected i.e. mammary glands in pigs, lungs in sheep, and dogs.[12] The disease apparently is rare in marine mammals. A thoracic actinomycosis has been reported in a seal from a zoological garden in Frankfurt, Germany.[75] Sweeney et al.[5] in their review of systemic mycoses list *A. bovis* from a beached adult Atlantic bottle-nosed dolphin (*Tursiops truncatus*).

XVI. BOTRYOMYCOSIS

Botryomycosis is caused by a nonfilamentous bacterium. It is a chronic bacterial infection of the skin resulting in granulomatous abscesses which may spread to other parts of the body.[17,20] The centers of the abscesses contain granules. Within the granules are large colonies of bacteria, closely grouped together and surrounded by eosinophilic amorphous radiating clubs (Splendore-Hoeppli phenomenon).[49] Although many genera of bacteria have been implicated, the most common ones are *Staphylococcus aureus* and *Pseudomonas aeruginosa*. Special bacterial stains such as Giemsa and Brown-Brenn Gram methods are helpful in identifying the etiologic agents in tissue sections.[26]

Botryomycosis has been reported in many species including cattle and pigs in which it caused mastitis. A report of the disease in marine mammals involves three harp seals (*Pagophilus groenlandicus*).[76] These animals had been in captivity and shared a common outdoor pool at Guelph, Ontario, Canada. All had cutaneous lesions, and two of the seals had disseminated lesions. Since staphylococci were not cultured from the pool, contamination from a human source was thought to be the source of infection as coagulase-positive staphylococci can be isolated from the skin and mucous membrane of the nasal tract of humans. Staphylococcal botryomycosis also has been reported in a lymph node of a cetacean; it was a concurrent finding in a captive Pacific bottle-nosed dolphin (*Tursiops gilli*) infected with *Nocardia brasiliensis* in Hawaii.[66]

ACKNOWLEDGMENTS

This work was supported in part by Public Health Service Grant No. RR00301-16 from the Division of Research Resources, National Institutes of Health, U.S. Department of Health and Human Services, under the auspices of Universities Associated for Research and Education in Pathology, Inc. The authors thank all of the individuals who have contributed specimens from marine mammals for consultation to the Registry of Comparative Pathology, as much of the information presented is based on the examination of that material. We also thank individuals who have shared their interesting and unusual cases of mycotic infections with us.

The authors thank Merryanna Swartz for assistance in the literature search and Charmaine Goetz and Susan Allen for technical assistance.

The opinions or assertions contained herein are the private views of the authors and are not to be construed as official or as reflecting the views of the Department of the Army, the Department of the Air Force, or the Department of Defense.

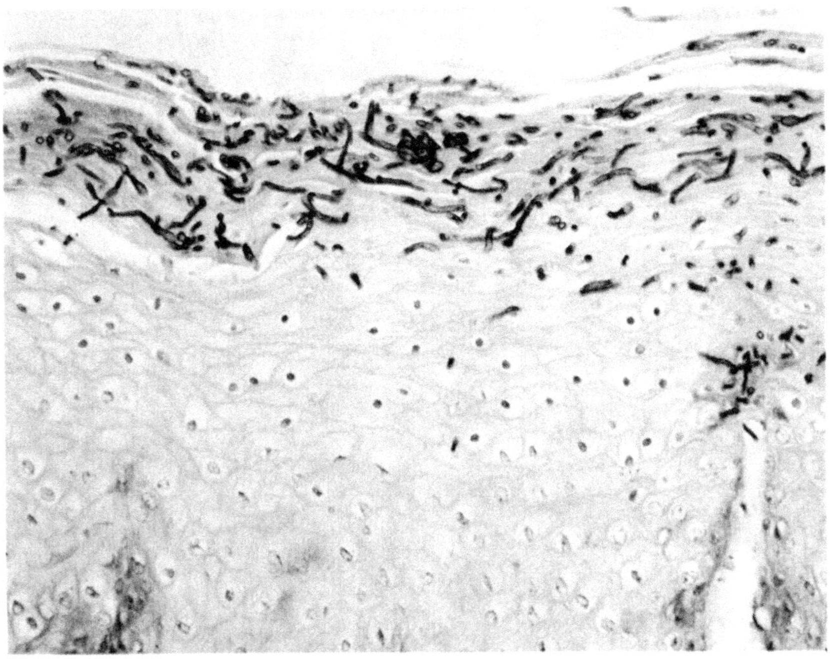

FIGURE 1. Esophageal candidiasis in a dolphin. PAS; ×250 (AFIP neg 82-5497).

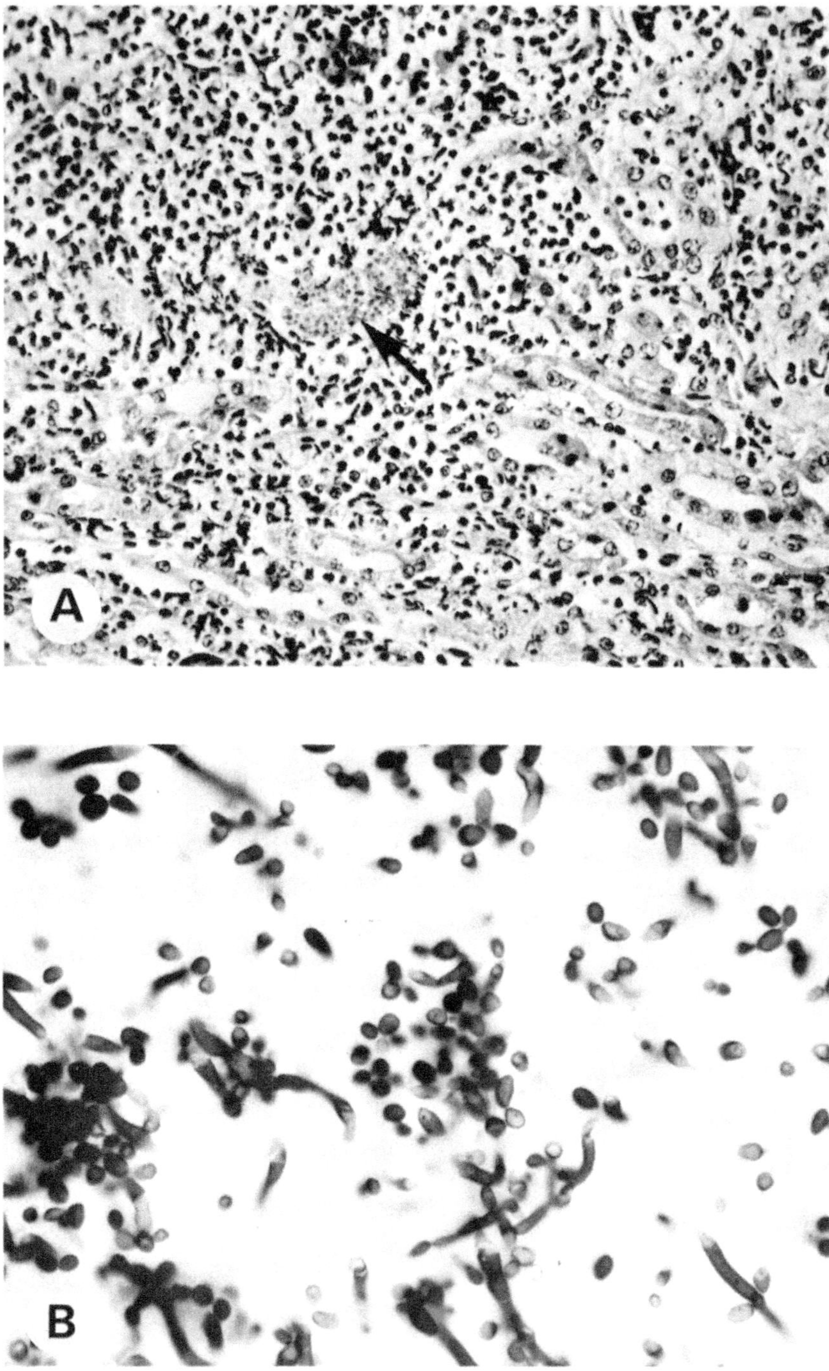

FIGURE 2. A. Necrosis and abscesses caused by *Candida* (arrow) in the kidney of a killer whale. H&E; × 250 (AFIP neg. 82-5504). B. GMS stained section to demonstrate large colonies of hyphae, pseudohyphae and blastospores. × 630 (AFIP neg 82-7195).

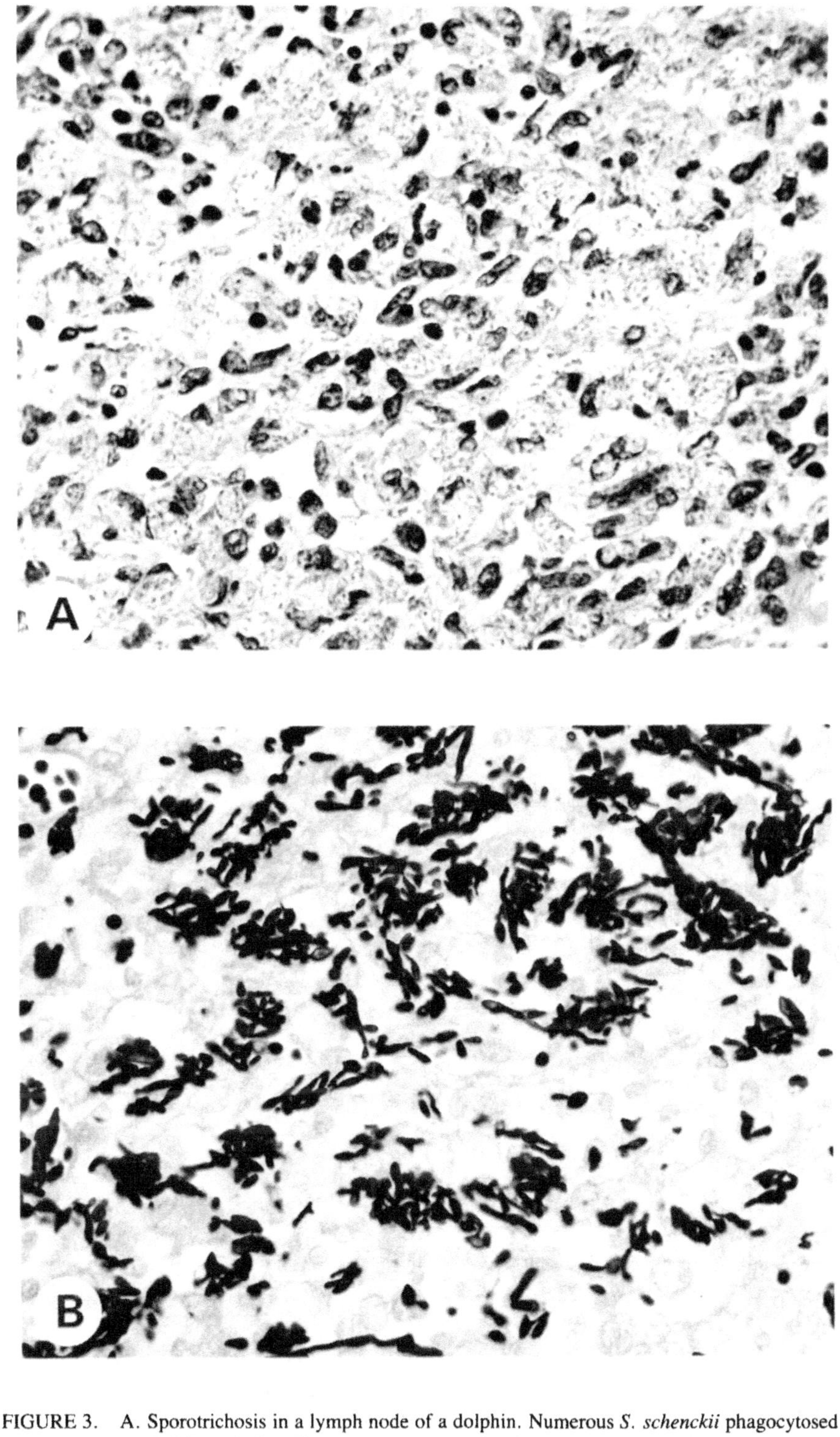

FIGURE 3. A. Sporotrichosis in a lymph node of a dolphin. Numerous *S. schenckii* phagocytosed by macrophages. H&E; ×525 (AFIP neg 78-554). B. GMS stained section to demonstrate ovoid to cigar-shaped characteristics. ×630 (AFIP neg 82-5516).

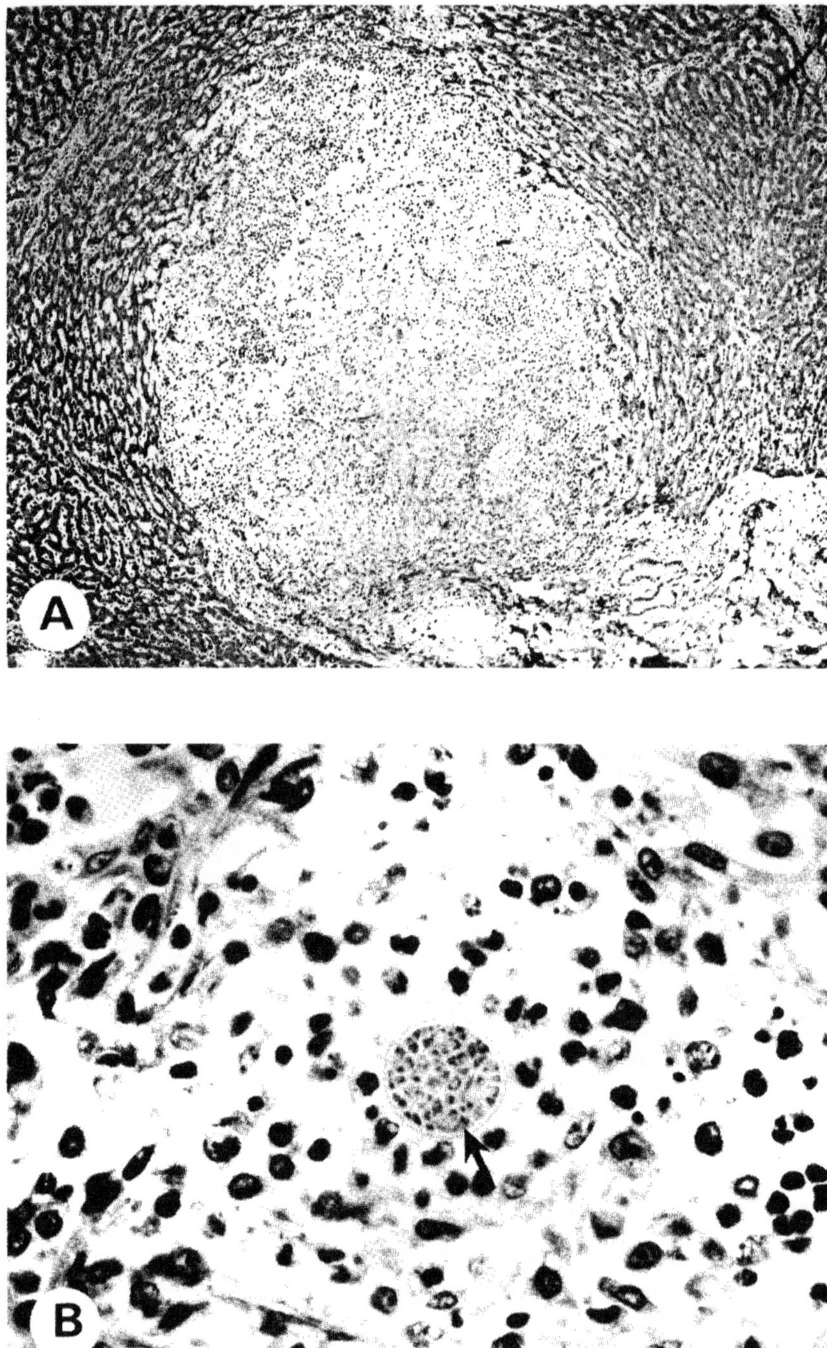

FIGURE 4. A. Coccidioidomycosis. Pyogranulomatous lesion in the liver of a sea lion. H&E; ×60 (AFIP neg 82-5513). B. Higher magnification to demonstrate a spherule (arrow) filled with endospores. H&E; ×630 (AFIP neg 82-7196).

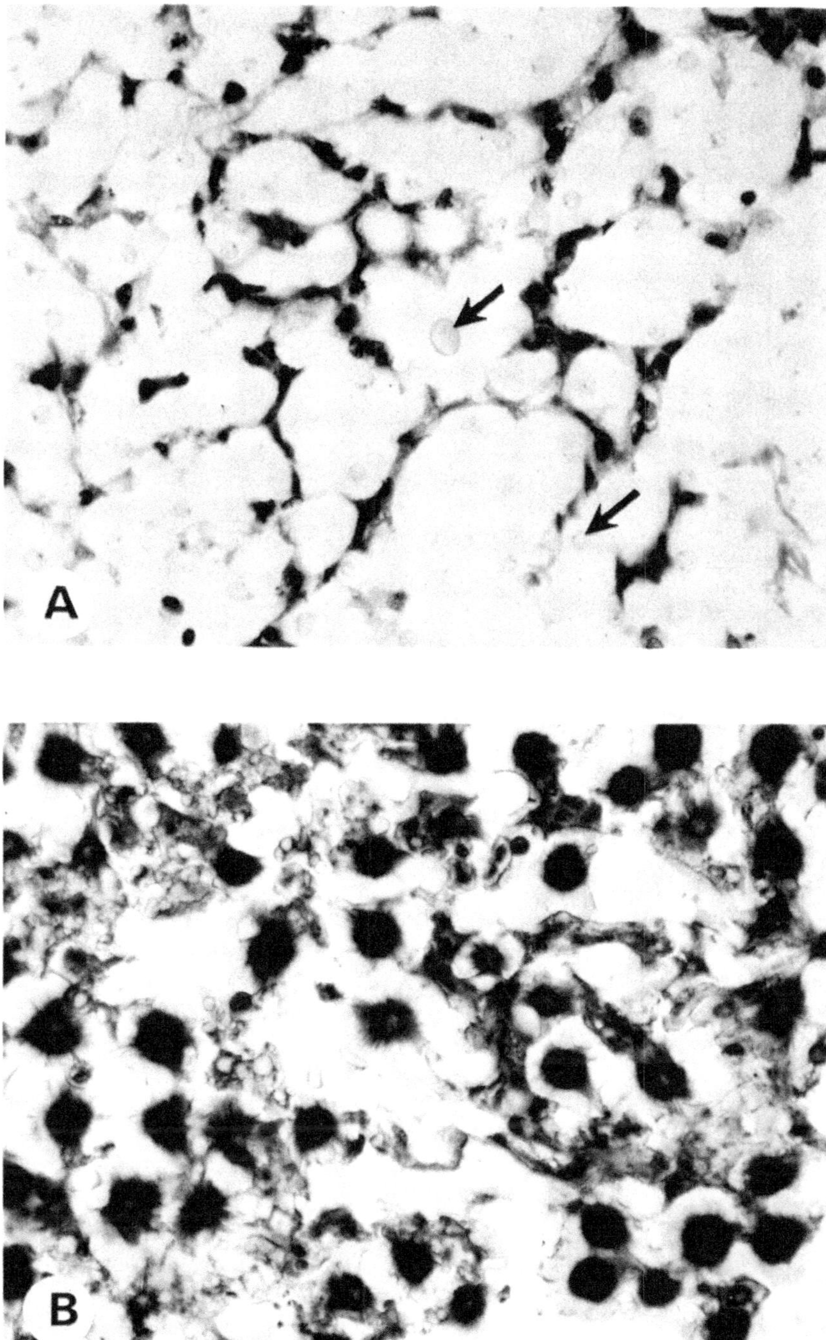

FIGURE 5. A. Cryptococcal pneumonitis in a dolphin; note wide clear spaces surrounding fungal cells (arrows). H&E; ×400 (AFIP neg 82-7204). B. Mayer's mucicarmine stained section to demonstrate the mucinous capsule. ×400 (AFIP neg 82-7207).

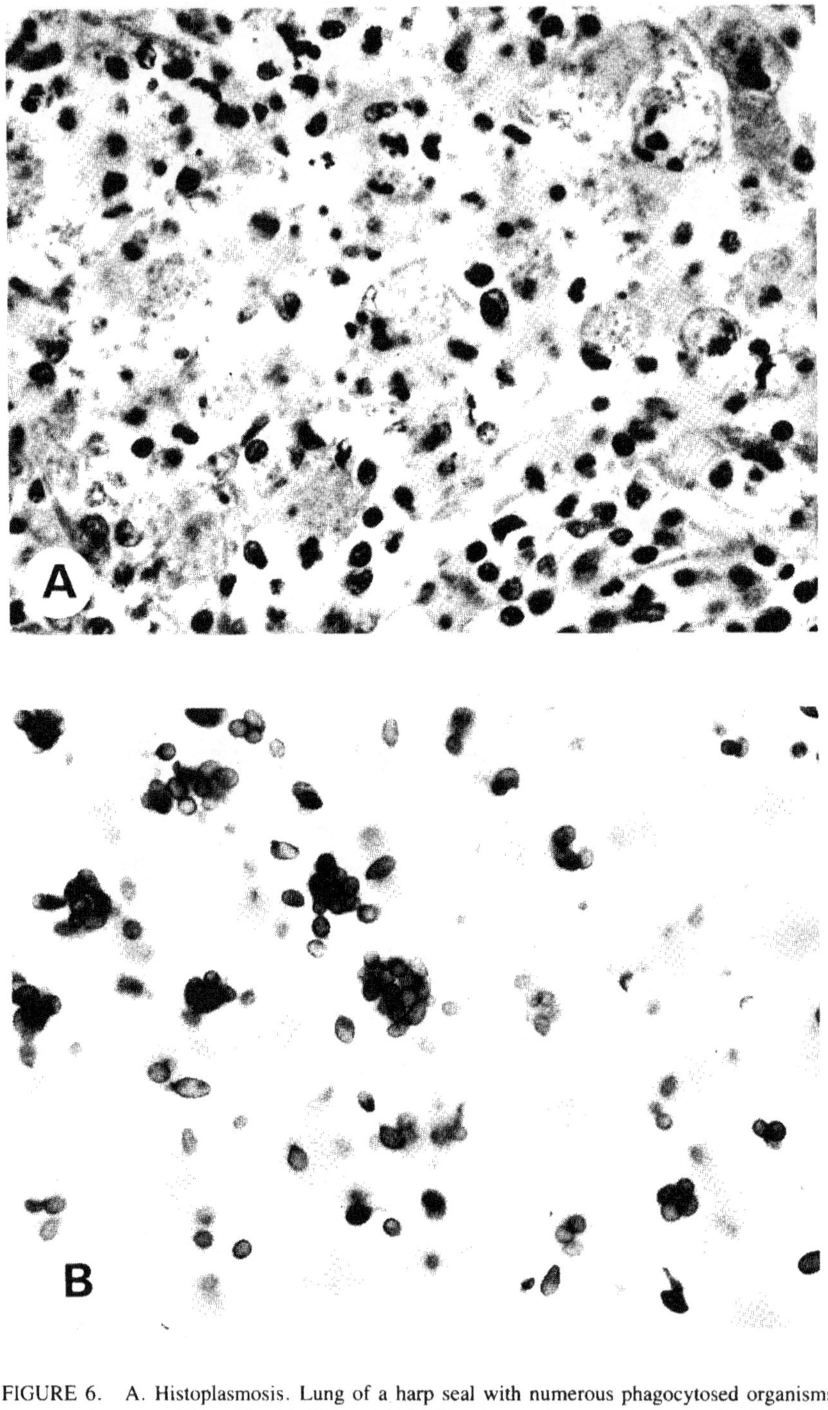

FIGURE 6. A. Histoplasmosis. Lung of a harp seal with numerous phagocytosed organisms in macrophages. H&E; ×630 (AFIP neg 82-7198). B. Duplicate section stained with GMS. ×1000 (AFIP neg 82-7199).

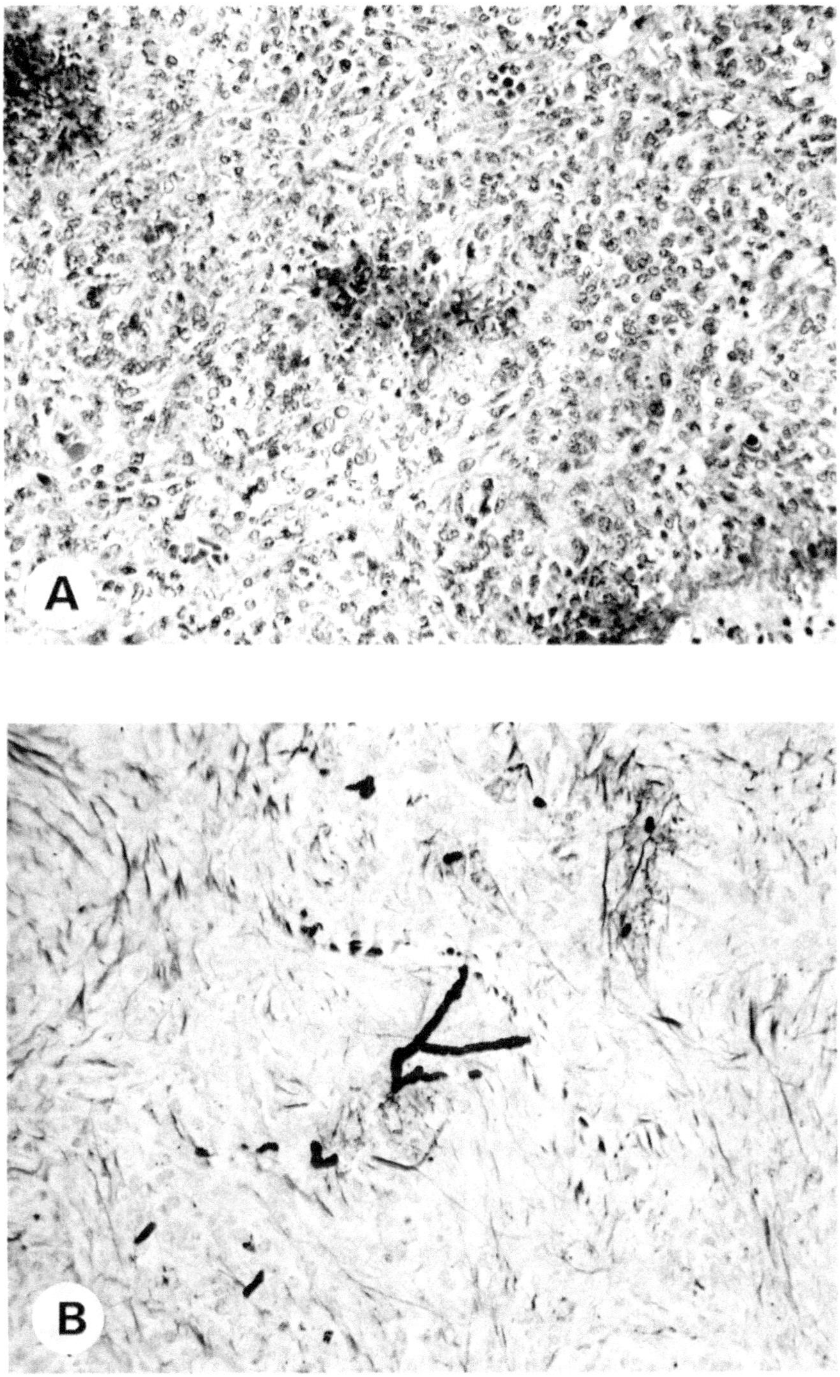

FIGURE 7. A. Aspergillosis. Chronic necrotic granulomatous lesions in the trachea of a dolphin. H&E; ×250 (AFIP neg 82-5508). B. GMS stained section to demonstrate septate hyphae and dichotomous branching. ×250 (AFIP neg 82-5509).

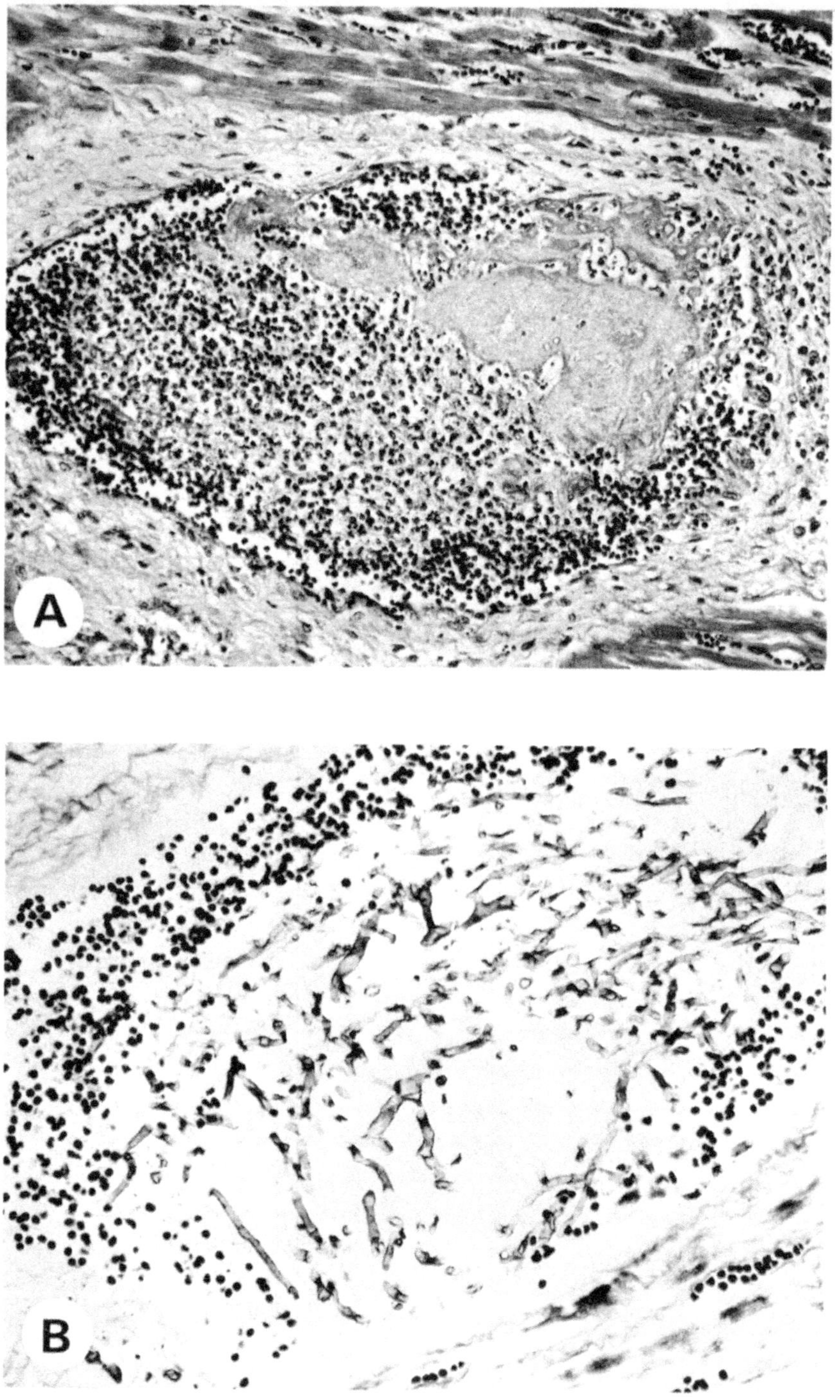

FIGURE 8. A. Zygomycosis. Mycotic thrombi in the heart of a killer whale. H&E; × 60 (AFIP neg 82-5517). B. GMS stained section to demonstrate hyphal organisms in blood vessel. ×250 (AFIP neg 82-5519).

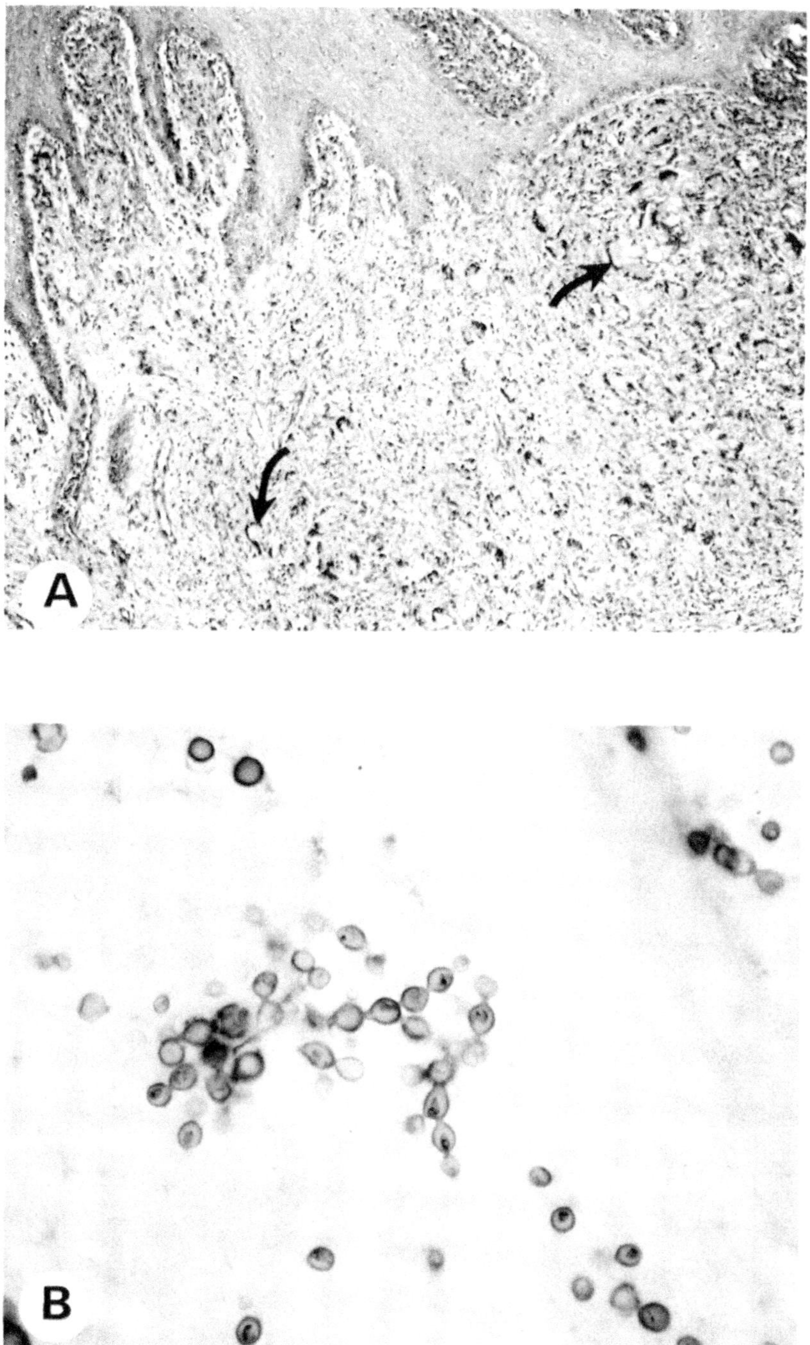

FIGURE 9. A. Lobomycosis. Histiocytic granuloma in the dermis of a dolphin. Numerous multinucleated giant cells (arrows) are present. H&E; ×60 (AFIP neg 82-5510). B. Gridley stained section to demonstrate branching yeast forms of *Loboa loboi*. × 1000 (AFIP neg 82-7203).

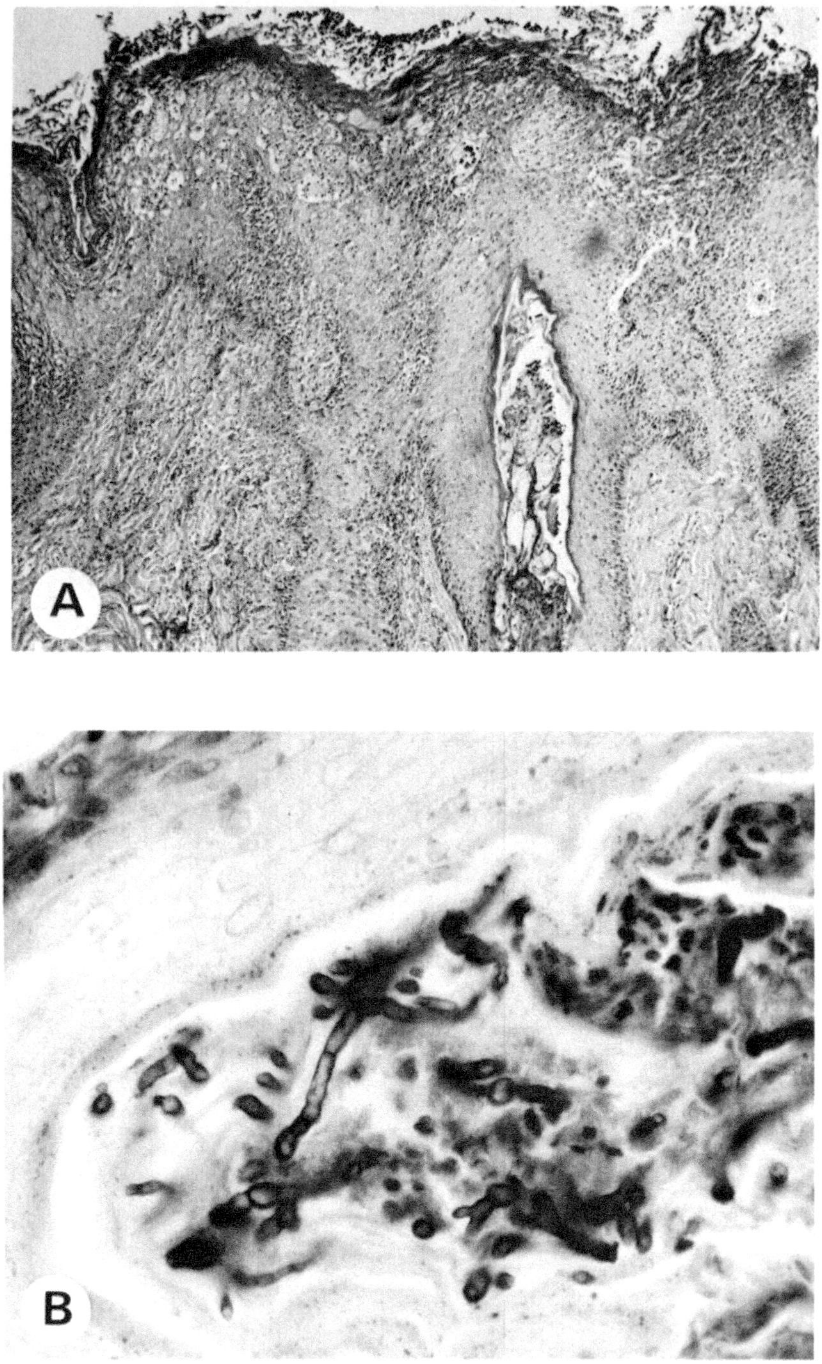

FIGURE 10. A. Fusariomycosis. Hyperkeratosis, acanthosis and intraepithelial abscesses in the epidermis of a sea lion. H&E; ×60 (AFIP neg 82-5491). B. PAS stained section to demonstrate septate branching hyphae. ×630 (AFIP neg 82-7201).

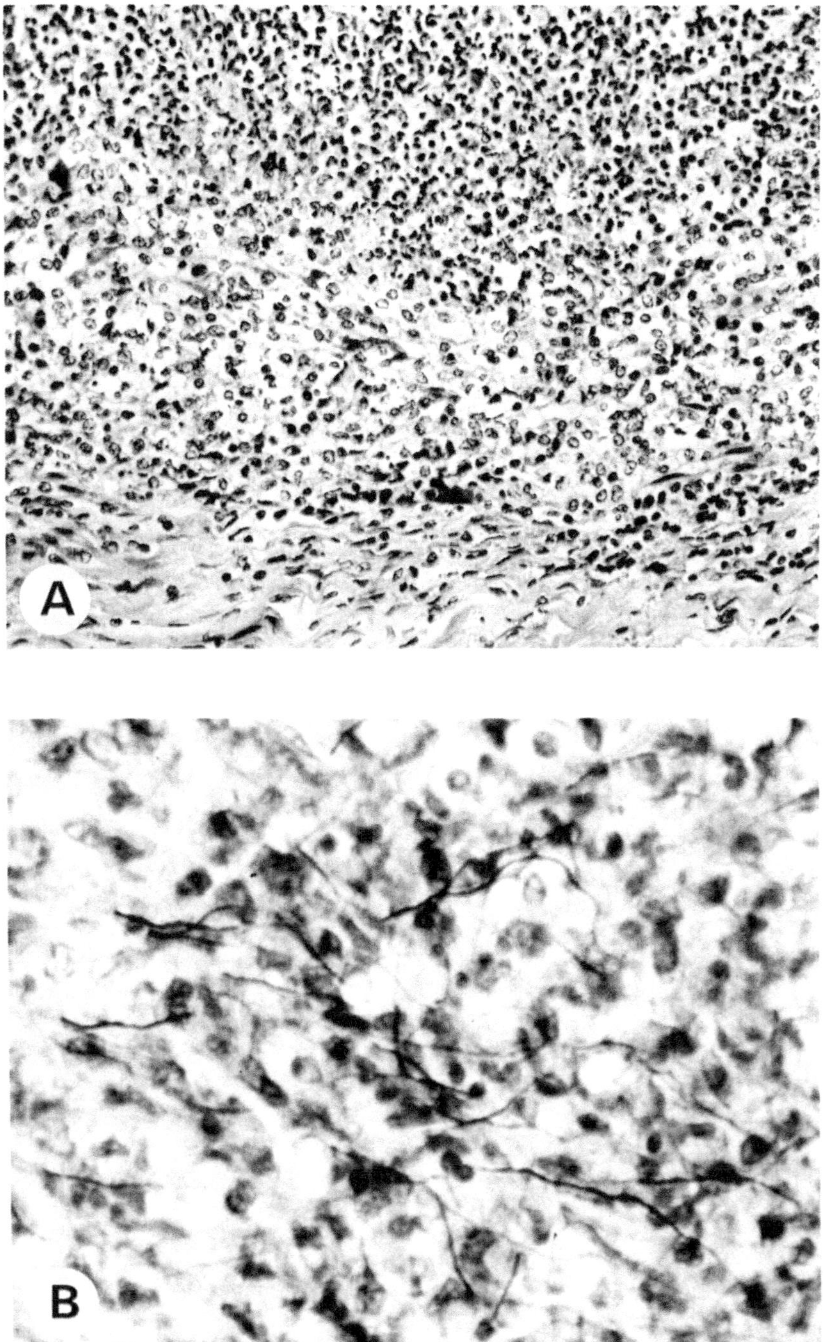

FIGURE 11. A. Nocardiosis. Chronic pyogranulomatous pleuritis in a Beluga whale. H&E; ×250 (AFIP neg 82-5499). B. Fite-Faraco acid-fast stained section to demonstrate branching filamentous organisms. ×1040 (AFIP neg 82-5500).

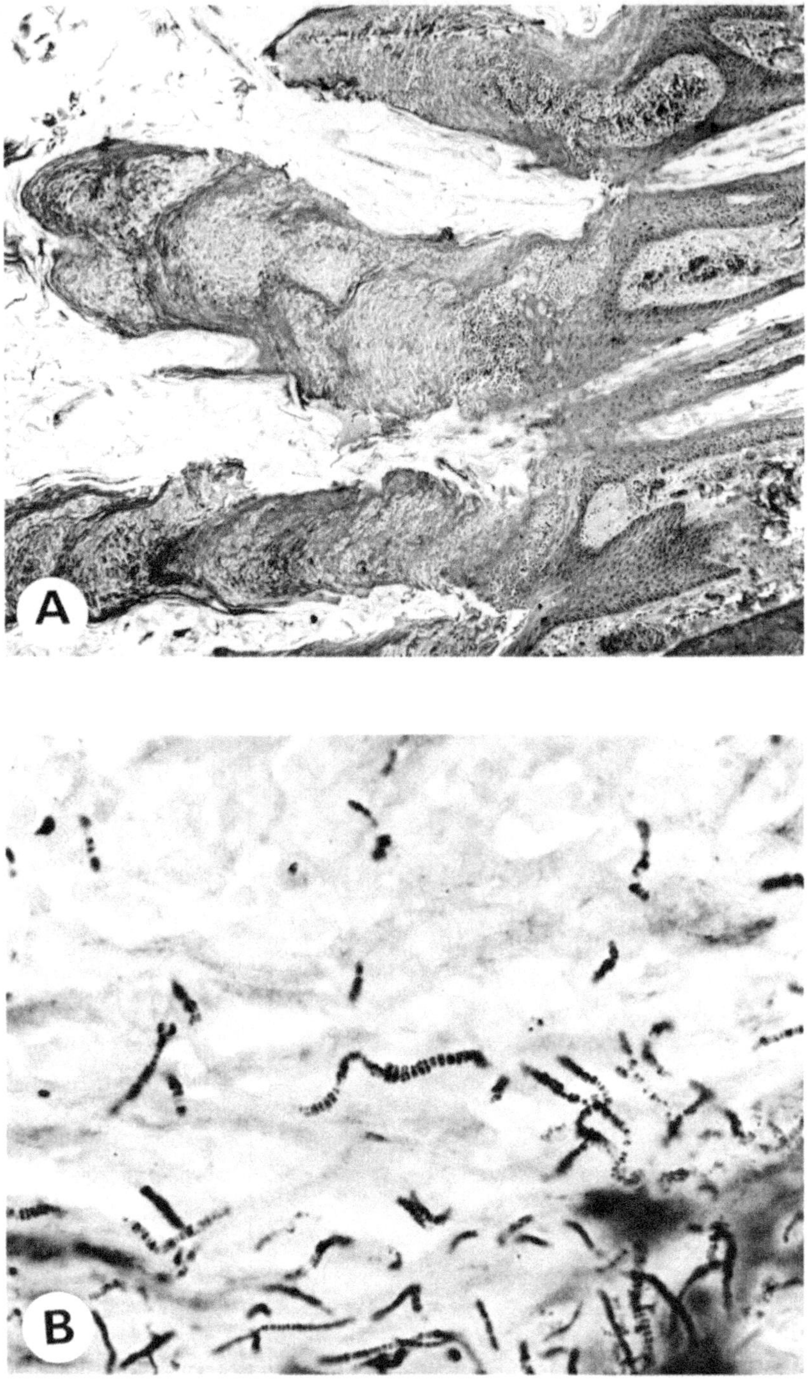

FIGURE 12. A. Dermatophilosis in a sea lion. Note alternate layers of suppurative exudate and cornified epidermis. H&E; ×60 (AFIP neg 82-5494). B. Brown-Brenn Gram stained section to demonstrate multidimensional division of *D. congolensis.* ×1000 (AFIP 82-5407).

REFERENCES

1. **Sweeney, J. C. and Ridgway, S. H.,** Common diseases of small cetaceans, *J. Am. Vet. Med. Assoc.*, 167, 533, 1975.
2. **Sweeney, J. C.,** Common diseases of pinnipeds, *J. Am. Vet. Med. Assoc.*, 165, 805, 1974.
3. **Ridgway, S. H.,** *Mammals of the Sea—Biology and Medicine,* Charles C Thomas, Springfield, Ill., 1972.
4. **Geraci, J. R.,** Marine mammals (Cetacea, Pinnipedia and Sirenia), in *Zoo and Wild Animal Medicine,* Fowler, M. E., Ed., W. B. Saunders, Philadelphia, 1978, chap. 25.
5. **Sweeney, J. C., Migaki, G., Vainik, P. M., and Conklin, R. H.,** Systemic mycoses in marine mammals, *J. Am. Vet. Med. Assoc.*, 169, 946, 1976.
6. **Medway, W.,** Some bacterial and mycotic diseases of marine mammals, *J. Am. Vet. Med. Assoc.*, 177, 831, 1980.
7. **Emmons, C. W.,** Natural occurrence of opportunistic fungi, *Lab. Invest.*, 11, 1026, 1962.
8. **Kaplan, W.,** Epidemiology of the principal systemic mycoses of man and lower animals and the ecology of their etiologic agents, *J. Am. Vet. Med. Assoc.*, 163, 1043, 1973.
9. **Connole, M. D. and Johnston, L. A. Y.,** A review of animal mycoses in Australia, *Vet. Bull.*, 37, 145, 1967.
10. **Emmons, C. W., Binford, C. H., Utz, J. P., and Kwon-Chung, K. J.,** *Medical Mycology,* 3rd ed., Lea and Febiger, Philadelphia, 1977.
11. **Migaki, G.,** Mycotic diseases in captive animals — a mycopathologic overview, in *The Comparative Pathology of Zoo Animals,* Montali, R. J. and Migaki, G., Eds., Smithsonian Institution Press, Washington, D.C., 1980, 267.
12. **Ainsworth, G. C. and Austwick, P. K. C.,** *Fungal Diseases of Animals,* 2nd ed., Commonwealth Agricultural Bureaux, Farnham Royal, Slough, England, 1973.
13. **Bridges, C. H.,** Fungous Diseases, in *Diseases Transmitted from Animals to Man,* 5th ed., Hull, T. G., Ed., Charles C Thomas, Springfield, Ill., 1963, 453.
14. **Smith, J. M. B.,** Diseases of laboratory animals — mycotic, in *Handbook of Laboratory Animal Science,* Vol. 2, Melby, E. C. Jr. and Altman, N. H., Eds., CRC Press, Inc., Boca Raton, Fla., 1974, 331.
15. **Al-Doory, Y.,** Fungal and bacterial diseases, in *Pathology of Simian Primates,* Part II, Fiennes, R. N. T-W., Ed., S. Karger, Basel, 1972, 206.
16. **Jungerman, P. F. and Schwartzman, R. M.,** *Veterinary Medical Mycology,* Lea and Febiger, Philadelphia, 1972.
17. **Chandler, F. W., Kaplan, W., and Ajello, L.,** *Histopathology of Mycotic Diseases,* Year Book Medical Publishers, Inc., Chicago, Ill., 1980.
18. **Migaki, G., Voelker, F. A., and Sagartz, J. W.,** Fungal diseases, in *Pathology of Laboratory Animals,* Vol. 2, Benirschke, K., Garner, F. M., and Jones, T. C., Eds., Springer-Verlag, New York, 1978, 1551.
19. **Baker, R. D.,** *Human Infection with Fungi, Actinomycetes and Algae,* Springer-Verlag, New York, 1971.
20. **Rippon, J. W.,** *Medical Mycology,* W. B. Saunders, Philadelphia, 1974.
21. **Binford, C. H. and Dooley, J. R.,** Diseases caused by fungi and actinomycetes, in *Pathology of Tropical and Extraordinary Diseases,* Vol. 2, Binford, C. H. and Connor, D. H., Eds., Armed Forces Institute of Pathology, Washington, D.C., 1976, 551.
22. **Conant, N. F., Smith, D. T., Baker, R. D., and Callaway, J. L.,** *Manual of Clinical Mycology,* 3rd ed., W. B. Saunders, Philadelphia, 1971.
23. **Carter, G. R.,** *Diagnostic Procedures in Veterinary Bacteriology and Mycology,* 3rd ed., Charles C Thomas, Springfield, Ill., 1979.
24. **Ajello, L.,** Comments on the laboratory diagnosis of opportunistic fungus diseases, *Lab. Invest.*, 11, 1033, 1962.
25. **Kaplan, W. and Kraft, D. E.,** Demonstration of pathogenic fungi in formalin-fixed tissues by immunofluorescence, *Am. J. Clin. Pathol.*, 52, 420, 1969.
26. **Luna, L. G.,** *Manual of Histologic Staining Methods of the Armed Forces Institute of Pathology,* 3rd ed., McGraw-Hill, New York, 1968.
27. **Nakagawa, S., Masui, M., Tashiro, K., and Tanabe, K.,** A case of candidiasis in a harbor seal *(Phoca vitulina), J. Jpn. Ass. Zool. Gard. Aquat.*, 9, 95, 1967.
28. **Reichel, K. and Mayer, H.,** Candidiasis in seals *(Mirounga leonina), Prakt. Tierarzt.*, 59, 110, 1978.
29. **Ostenrath, F.,** Some remarks on therapy of mycotic and bacteriological skin diseases in freshwater dolphins *(Inia geoffrensis), Aquatic Mammals,* 4, 49, 1976.
30. **Nakeeb, S., Targowski, S. P., and Spotte, S.,** Chronic cutaneous candidiasis in bottle-nosed dolphins, *J. Am. Vet. Med. Assoc.*, 171, 961, 1977.
31. **Greenwood, A. G., Taylor, D. C., and Wild, D.,** Fibreoptic gastroscopy in dolphins, *Vet. Rec.*, 102, 495, 1978.
32. **Ridgway, S. H.,** Reported causes of death of captive killer whales *(Orcinus orca), J. Wildl. Dis.*, 15, 99, 1979.

33. **Finn, J. P.,** Pyocephalus and gastritis in a polar bear *(Thalarctus maritimus), J. Am. Vet. Med. Assoc.,* 155, 1086, 1969.
34. **Kier, A. B., Mann, P. C., and Wagner, J. E.,** Disseminated sporotrichosis in a cat, *J. Am. Vet. Med. Assoc.,* 175, 202, 1979.
35. **Migaki, G., Font, R. L., Kaplan, W., and Asper, E. D.,** Sporotrichosis in a Pacific white-sided dolphin *(Lagenorhynchus obliquidens), Am. J. Vet. Res.,* 39, 1916, 1978.
36. **Maddy, K. T.,** Coccidioidomycosis in animals, *Vet. Med.,* 54, 233, 1959.
37. **Reed, R. E., Migaki, G., and Cummings, J. A.,** Coccidioidomycosis in a California sea lion, *J. Wildlife Dis.,* 12, 372, 1976.
38. **Cornell, L. H., Osborn, K. G., Antrim, J. E., and Simpson, J. G.,** Coccidioidomycosis in a California sea otter, *J. Wildlife Dis.,* 15, 373, 1979.
39. **Furcolow, M. L., Chick, E. W., Busey, J. F., and Menges, R. W.,** Prevalence and incidence studies of human and canine blastomycosis. I. Cases in the United States, 1885-1968, *Am. Rev. Respir. Dis.,* 102, 60, 1970.
40. **Williamson, W. M., Lombard, L. S., and Getty, R. E.,** North American blastomycosis in a northern sea lion, *J. Am. Vet. Med. Assoc.,* 135, 513, 1959.
41. **Emmons, C. W.,** Saprophytic sources of *Cryptococcus neoformans* associated with the pigeon *(Columba livia), Am. J. Hyg.,* 62, 227, 1955.
42. **Griner, L. A. and Walch, H. A.,** Cryptococcosis in columbiformes at the San Diego Zoo, *J. Wildlife Dis.,* 14, 389, 1978.
43. **Pounden, W. D., Amberson, J. M., and Jaeger, R. F.,** A severe mastitis problem associated with *Cryptococcus neoformans* in a large dairy herd, *Am. J. Vet. Res.,* 13, 121, 1952.
44. **Barron, C. N.,** Cryptococcosis in animals, *J. Am. Vet. Med. Assoc.,* 127, 125, 1955.
45. **Migaki, G., Gunnels, R. D., and Casey, H. W.,** Pulmonary cryptococcosis in an Atlantic bottle-nosed dolphin *(Tursiops truncatus), Lab. Anim. Sci.,* 28, 603, 1978.
46. **Ajello, L.,** Comparative ecology of respiratory mycotic disease agents, *Bacteriol. Rev.,* 31, 6, 1967.
47. **Menges, R. W., Habermann, R. T., Selby, L. A., Ellis, H. R., Behlow, R. F., and Smith, C. D.,** A review and recent findings on histoplasmosis in animals, *Vet. Med.,* 58, 331, 1963.
48. **Wilson, T. M., Kierstead, M., and Long, J. R.,** Histoplasmosis in a harp seal, *J. Am. Vet. Med. Assoc.,* 165, 815, 1974.
49. **Johnson, F. B.,** Splendore-Hoeppli phenomenon, in *Pathology of Tropical and Extraordinary Diseases,* Vol. 2, Binford, C. H. and Connor, D. H., Eds., Armed Forces Institute of Pathology, Washington 1976, 681.
50. **Carroll, J. M., Jasmin, A. M., and Baucom, J. N.,** Pulmonary aspergillosis of the bottle-nose dolphin *(Tursiops truncatus), Am. J. Vet. Clin. Pathol.,* 2, 139, 1968.
51. **Kaplan, W., Goss, L. J., Ajello, L., and Ivens, M. S.,** Pulmonary mucormycosis in a harp seal caused by *Mucor pusillus, Mycopathol. Mycol. Appl.,* 12, 101, 1960.
52. **Best, P. B. and McCully, R. M.,** Zygomycosis (Phycomycosis) in a right whale *(Eubalaena australis), J. Comp. Pathol.,* 89, 341, 1979.
53. **Migaki, G. and Blumer, P. W.,** Case for diagnosis — phycomycosis, *Mil. Med.,* 140, 544, 1975.
54. **Wiersema, J. P. and Niemel, P. L. A.,** Lobo's disease in Surinam patients, *Trop. Geogr. Med.,* 17, 89, 1965.
55. **Migaki, G., Valerio, M. G., Irvine, B., and Garner, F. M.,** Lobo's disease in an Atlantic bottle-nosed dolphin, *J. Am. Vet. Med. Assoc.,* 159, 578, 1971.
56. **Woodard, J. C.,** Electron microscopic study of lobomycosis *(Loboa loboi), Lab. Invest.,* 27, 606, 1972.
57. **de Vries, G. A. and Laarman, J. J.,** A case of Lobo's disease in the dolphin *Sotalia guianensis, Aquatic Mammals,* 1, 26, 1973.
58. **Poelma, F. G., de Vries, G. A., Blythe-Russell, E. A., and Luykx, M. H. F.,** Lobomycosis in an Atlantic bottle-nosed dolphin in the Dolphinarium Harderwijk, *Aquatic Mammals,* 2, 11, 1974.
59. **Goldston, R. T., Whitman, R. A., and Haslett, T. E.,** Lobo's disease in the Atlantic bottle-nose dolphin (*Tursiops truncatus),* in *Proc. Am. Assoc. Zoo Vet.,* Atlanta, 1974, 128.
60. **Caldwell, D. K., Caldwell, M. C., Woodard, J. C., Ajello, L., Kaplan, W., and McClure, H. M.,** Lobomycosis as a disease of the Atlantic bottle-nosed dolphin *(Tursiops truncatus* Montagu, 1821), *Am. J. Trop. Med. Hyg.,* 24, 105, 1975.
61. **Jacobson, E. R.,** Mycotic diseases of reptiles, in *The Comparative Pathology of Zoo Animals,* Montali, R. J. and Migaki, G., Eds., Smithsonian Institution Press, Washington, D.C., 1980, 283.
62. **Montali, R. J., Bush, M., Strandberg, J. D., Janssen, D. L., Boness, D. J., and Whitla, J. C.,** Cyclic dermatitis associated with *Fusarium* sp infection in pinnipeds, *J. Am. Vet. Med. Assoc.,* 179, 1198, 1981.
63. **Hoshina, T. and Sugiura, Y.,** On a skin disease and a nematode parasite of a dolphin *Tursiops truncatus* (Montagu, 1821), *Sci. Rep. Whales Res. Inst. (Tokyo),* 11, 133, 1956.
64. **Dilbone, R. P.,** Mycosis in a manatee, *J. Am. Vet. Med. Assoc.,* 147, 1095, 1965.

65. **Farnsworth, R. J., McKeever, P. J., and Fletcher, J. A.,** Dermatomycosis in a harbor seal caused by *Microsporum canis*, *J. Zoo Anim. Med.*, 6, 26, 1975.
66. **Pier, A. C., Takayama, A. K., and Miyahara, A. Y.,** Cetacean nocardiosis, *J. Wildlife Dis.*, 6, 112, 1970.
67. **MacNeill, A. C., Gornall, T. A., Giddens, W. E., and Boyce, J.,** Evidence of *Nocardia* sp. in a captive-born Beluga whale, *Aquatic Mammals*, 8, 50, 1980.
68. **Jasmin, A. M., Powell, C. P., and Baucom, J. N.,** Actinomycotic mycetoma in the bottle-nose dolphin (*Tursiops truncatus*) due to *Nocardia paraguayensis*, *Vet. Med. Small Anim. Clin.*, 67, 542, 1972.
69. **Davis, G. B., Stevenson, B. J., Kyle, R. J., and Price, M. C.,** Isolation of an actinomycete from a sea leopard *(Hydrurga leptonyx)*, *N. Z. Vet. J.*, 25, 274, 1977.
70. **Roberts, D. S.,** Dermatophilus infection, *Vet. Bull.*, 37, 513, 1967.
71. **Montali, R. J., Smith, E. E., Davenport, M., and Bush, M.,** Dermatophilosis in Australian bearded lizards, *J. Am. Vet. Med. Assoc.*, 167, 553, 1975.
72. **Smith, C. F. and Cordes, D. O.,** Dermatitis caused by *Dermatophilus congolensis* infection in polar bears *(Thalactos maritimus)*, *Br. Vet. J.*, 128, 366, 1972.
73. **Newman, M. S., Cook, R. W., Appelhof, W. K., and Kitchen, H.,** Dermatophilosis in two polar bears, *J. Am. Vet. Med. Assoc.*, 167, 561, 1975.
74. **Frese, K. and Weber, A.,** Dermatitis in seals (*Otaria bryonia*, Blainville) caused by *Dermatophilus congolensis*, *Berl. Muench. Tieraerztl. Wochenschr.*, 84, 50, 1971.
75. **Schmitt, J.,** Aktinomykose bei einer Mähnenrobbe, *Nord. Veterinaermed.*, 14, 170, 1962.
76. **Wilson, T. M. and Long, J. R.,** The harp seal, *Pagophilus groenlandicus* (Erxleben, 1777) XII. Staphylococcal granulomas (botryomycosis) in harp seals, *J. Wildlife Dis.*, 6, 155, 1970.

Chapter 2

DISEASES ASSOCIATED WITH STRANDING AND CAPTIVITY

John G. Simpson and Lanny H. Cornell

TABLE OF CONTENTS

I. INTRODUCTION

It would be nice to present this chapter as the definitive and complete compendium of diseases of marine mammals related to stranding and captivity, but one must admit that a treatise of those dimensions will probably never be accomplished. The reason for this rather gloomy outlook is that the documentation of disease in these species is limited by the infinitely few cases which are available for examination. Death at sea, or on shores far from human access will undoubtedly continue to claim the vast majority of marine mammals decedents.

There are other problems associated with gathering accurate information and these include the fact that many times the necropsy of these animals is performed by interested, but medically untrained personnel, and thus detail may be overlooked, and specimens that are submitted for pathologic examination may not be representative, and except for a few instances, the study of marine mammal disease is not approached as an experimental or organized study, but is essentially an ad-hoc situation. Obviously, this results primarily from a lack of significant funding for this type of research. Thus information is provided by the interested individuals who have access to the animals as they happen to become ill or die on beaches, or in zoos and oceanariums.

The U.S. Navy Department, in the 1960s and 1970s provided a substantial amount of information related to physiology, anatomy, and diseases of marine mammals. Some oceanariums have added veterinarians to their staff, and a significant portion of the current knowledge regarding marine mammals is a result of the interaction between these individuals and various government and university laboratories.

We should not be surprised at the occurrence of any disease found in marine mammals, in that over a period of time, diseases of all classes have been seen in these animals. Thus, we find in this book, references to neoplastic, infectious, and metabolic conditions. The major question is how many of these diseases are found in feral conditions, and which ones are a product of contact with the terrestrial environment in zoos and oceanaria. In some instances, the fact that a specific animal is host to a certain class of disease is not as interesting as the finding of a particular disease in an unlikely environment, such as coccidioidomycosis in a feral sea otter, since this condition (''Valley Fever'') is generally associated with an arid climate.[1]

One major source of disease information is that derived from stranded animals, either pinnipeds or cetaceans. There is a significant difference in public interest related to strandings, and thus varying probability that the animal will be brought to the attention of a marine mammal scientist. Seals or sea lions commonly are found individually deteriorating on the beach or areas of the shore (Figure 1). These occurrences do not generally create the media that cetacean strandings seem to generate. Any concern of the public about pinniped strandings usually involves someone rescuing a seal or sea lion and then attempting to nurse it back to health. Recovery is sometimes accomplished, particularly with young animals, because the reason for distress is more often malnourishment or abandonment by the mother. In older animals, the success rate is much diminished, because the reason for beaching is more likely an infectious, neoplastic or degenerative condition from which the animal is less likely to survive. In California, a common cause of strandings in California sea lions is pneumonia, caused by the helminthic parasite, *P. decorus,* and the intermediate host is the opal-eye fish (*G. nigricans*).[2] On the east coast another nematode parasite, *O. circumlitus,* causes similar problems in Harbor seals.[3] Examples of viral and bacterial related strandings have been reported, including a hepatitis caused by an adenovirus,[4] and a nephritis associated with *Leptospira* sp. in California sea lions.[5]

Cetacean strandings are less common and usually cause much more media attention, evoking a rash of hypotheses concerning the cause of the stranding, ranging from possibly correct, to rather fanciful. Certainly one reason for the difference between cetacean strandings and those of pinnipeds, is that when porpoises or whales come ashore and are discovered by humans, it is a striking and somewhat uncommon event. Seals and sea lions are more commonly seen in littoral areas, and do not seem as unusual when observed on a beach.

An implication could be made, that some atavistic impulse triggered by a gallimaufry of possible etiologies is the motivating force responsible for strandings. Generally, mass cetacean strandings involve relatively few (under 50) animals, but occasionally, large numbers may come ashore. A stranding of 150 dolphins occurred on the Japanese island of Kyushu, in January of 1982. A marine life scientist speculated that the stranding was related to disruption of echo location, but no specific cause was proven. Death and strandings of pinnipeds, not associated with natural causes, may be caused by gunshot wounds, and in one report of a group of 86 marine mammals stranded along the Oregon coast, 30% died as a result of trauma from bullets (Figure 2).[6]

Most cetaceans stranded on a shoreline finally die from overheating, irregardless of why they came ashore in the first place. The larger the animal the more rapidly it overheats out of the water, since they are designed to be immersed, and air only has about 1/25th the capability to disperse heat as well as water does. Thus, even with their countercurrent circulatory ability, the animals' internal temperature, generally aided by solar heating, rises until the body temperature reaches heat prostration level, and cardiovascular collapse occurs. In some larger cetaceans, this type of overheating can occur in a few minutes. Often, no single cause of stranding or death can then be determined, especially as most tissues are autolyzed by the time the necropsy examination is performed.

There have been many theories regarding the cause for cetacean strandings, but little research of substance has evolved. The four general theories have related to disease, disturbance of echo location, pursuit of food, and escape from danger. Indeed, with both pinnipeds and cetaceans, credence must be given to the thought that a mammal will return to land when it is not able to sustain itself in a pelagic environment. Ridgway and Daily[7] have found parasites in the brain of dolphins, and speculated that this may disrupt normal echo location, thus resulting in beachings. However, the mere presence of a parasitic encephalopathy may constitute a debility in the animal, apart from echo location problems which causes stranding. In a group of Sperm whales which came ashore in Florida in 1980, the striking feature was a hepatopathy, characterized by hemorrhage, necrosis, and inflammation (Figure 3). No brain tissue was available so this parameter could not be assessed. It might be that any cetacean stranding involves only a portion of a herd and the stranded animals have a common disease. This would seem reasonable in that in any group of animals, it is rare to have an infectious condition that clinically affects all individuals of that group, and it would be reasonable to assume that other animals in the herd could undergo debilitation simultaneously in the case of parasitic infection, because of the relative uniform life cycle periods of most parasites. Wood[3] has proposed that cetaceans, under stress, revert to a primitive survival mode, that of seeking safety on land.

Marine mammals are acquired, whether it be for public display or research, from a variety of sources. These include a few private individuals who deal in procurement, strandings, zoological trades, and groups from government or private oceanariums or zoos who organize collecting expeditions. In capturing these species, neurogenic or capture shock, pre-existing disease, and occasional trauma are conditions which may immediately present themselves. The use of improper equipment can inflict skin lesions. So called "neurogenic" or capture shock presents as acute debilitation immediately after capture, or it may also be seen as a gradual deterioration with associated myopathy, followed by death, within a week or two of arrival.

The immediate deaths are usually attributable to neurogenic shock coupled with pre-existing disease, often parasitic. The more delayed deaths are often related to stress myopathy, which may be partly caused by lactic acid buildup in muscle tissue, with pH change and acute muscle bundle necrosis.

Pinnipeds seem to adapt to captivity in a more sanguine manner than do cetaceans. This is probably a result of the aquatic environment to which the cetacean is totally dependent, in contrast to the partially terrestrial existence of seals, sea lions, walruses, etc. Certainly, with cetaceans, overheating during transport can be more of a factor. It should be noted that the clinical symptoms of neurogenic or capture shock are not specific, and at necropsy, pneumonia or sepsis may have been the immediate cause of death. Pinnipeds may be shipped without undo discomfort for long distances, while porpoises and whales must be provided with special carriers to assure continual moisturization of the skin. In recent times, cetaceans, including rather large ones (e.g., killer whales), and pinnipeds are transported thousands of miles by jet aircraft and truck with no ill effects.

For reasons not yet understood, some species, particularly cetaceans, do not adjust readily, and may die irregardless of current therapeutic measures, shortly after capture. To illustrate this point, a variety of more exotic cetaceans can be obtained, but not sustained in captivity. Examples include the freshwater species *Inia geofrensis* (Amazon dolphin), *Platanista gangetica* (Ganges River dolphin), and marine animals such as *Phocoenoides* sp. Possibly we may someday know the secrets of survival in captivity for these animals. But it is improbable, because of space and diet requirements (plankton), that Baleen whales will ever be adapted to a captive environment.

II. WATER CONDITIONS

On a logical basis, water in which marine mammals are kept should be as pure as possible. Anyone who was working with these animals, particularly cetaceans, when they were first maintained for exhibition or research, can remember the infected skin lesions which were common when water in the holding pools was obviously impure. Standards set by the U.S. Government now suggest that the coliform count should not exceed 1000 colonies/100 mℓ of water.[8] These are fairly liberal standards. While not as important with pinnipeds, the salinity is of utmost significance in maintaining the integrity of cetacean skin. Ocean water is normally 3.4% salt by volume. It was demonstrated that if the volume drops below 1.0%, irregular necrosis and severe ulceration will occur in the epidermis.[9] Histologically, ballooning degeneration of the parakeratotic and prickle cell layers is first noted (Figure 4), followed by ulceration. The lesions usually start after the animal has been in the hypoosmolar water for approximately a week. After three weeks, the lesions are advanced. If the cetacean is placed back into normal sea water before lesions are irreversible (within 1 to 2 weeks), complete recovery of the integument can be expected.

Fortunately, pinnipeds do not seem to require sea water, and can adjust quite well to lightly chlorinated freshwater. Workers at the National Zoological Park in Washington, D.C., have found that fresh water chlorinated at approximately 0.5 ppm, and kept at 68 to 70° F (20 to 21° C), seems to be adequate for maintaining pinnipeds.[10] Simpson was surprised, when he visited Longleat Castle, near Bath, England, to see what appeared to be a California Sea Lion, swimming in a freshwater lake on the Castle grounds. The gentleman charged with caring for the rather large number of animals on the estate revealed that those were indeed California Sea Lions that had been obtained several years before from Santa Barbara, California and that they were existing quite nicely in their new, noble surroundings. Knott's Berry Farm in Los Angeles had California Sea Lions in freshwater for about 15 years before they learned marine mammals could not be kept in freshwater, and closed the exhibit.

III. NUTRITION

Nutrition is a matter of rather great concern with the initial adaptation to captivity and long-term maintenance of marine mammals species. The information concerning the natural diet in marine mammals has been achieved primarily from the examination of stomach contents of animals which have died in tuna nets, been stranded and died, or washed ashore dead. That seals at least, have a rather eclectic diet is found at necropsy where plaster, stones, money, etc. seem to be sitting incidentally within the gastric lumen (Figure 5). This, of course, is a phenomenon of seals which have been on display in aquarium settings. It would seem reasonable to assume that odontocetes and pinnipeds, in nature, subsist on varieties of fish, squid, and possible crustaceans. And because of migratory patterns of many species, and availability of food, diet will vary. In a captive situation, it becomes necessary to provide a diet which the animal will accept, and one which satisfies the nutrient requirements particular to the species. In most instances this involves trial, intuition, and theory. It obviously would be, in general, best to feed fresh fish or squid, but this is not always possible, and resorting to frozen sea food is usually the alternative. Freezing and conditions of storage may destroy certain essential nutrients. Thiamine, vitamin C, and vitamin E deficiencies (Figure 6), as well as diet related histamine toxicity, have been reported in marine mammals.[11] Whether vitamin deficiencies occur in feral marine mammals is speculative. One investigator has suggested that North Atlantic seals are deficient in thiamine.[11]

There is indeed an apparent difference between odontocetes and pinnipeds in regard to their preference for food. In early 1970, Simpson was involved in an attempt to develop a prepared food for both porpoises and pinnipeds. The obvious reasons for this were convenience and the availability of a standard, steady supply of food. The proposed requirements were acceptability by porpoises and pinnipeds, nutritional adequacy, easy storage, a reasonably long shelf-life, a physical form for the food which makes handling and feeding easy, and free of parasites and bacterial pathogens. One of the first attempts was to put a product in a sausage casing. This seemed logical in that the processing would be simple, and the shape of the hot dog was not too terribly unlike the shape of some fish. With this in mind, a trial using commercial, beef based hot dogs was used on porpoises and California sea lions. Although some sea lions would unenthusiastically accept these offerings, porpoises would, except for one old female, never ingest the hot dog (Figure 7). The exceptive creature, subsisted solely on a diet of hot dogs for three weeks, without obvious distress, and several years later gave birth to a calf (no relationship implied). Finding the success rate with sausages essentially nil, the next trial involved filling sausage casings with fish protein concentrate (FPC). The acceptance factor here was also minimal, both in porpoises and sea lions. The animals would mouth and play with the FPC sausages (particularly the porpoises), but would soon lose interest. Why neither species would accept the FPC is interesting, since the porpoises lack taste buds, while pinnipeds are like terrestrial mammals and have fully developed organs of taste.[9] Certainly, porpoises with their "nose" sitting on top of their head, and lacking olfactory brain lobes, don't smell the food, particularly when it is immersed in water. Thus it is an enigma how this discrimination or "taste" for food occurs. It is often noted, even when they are fed fish, a porpoise may reject one species over another.

After the FPC sausage trials, an attempt was made to prepare FPC in loaves, which could be cut to desired size. In addition to the FPC, the product contained kelp (for textural and nutritional purposes), oil, NaCl, anti-oxidants, and surfactants. The material was pasteurized and water added. Trials with this product resulted in mixed results, with reasonable acceptance by pinnipeds, but universal nonacceptance by porpoises. Subsequently, similar products were prepared commercially and used in some zoos, etc., as a pinniped ration. The use of these commercial foods has fallen into disuse currently. It seems that the bottom line in marine mammal feeding programs is that the most satisfactory diet consists of fresh fish (or squid to some extent), and secondarily, fresh frozen fish or squid (Figure 8). The most important consideration in using these sources is that they have been stored properly, and if possible, are free of parasites.

IV. ACCLIMATION TO CAPTIVITY

The first consideration in the maintenance of a marine mammal in captivity, is insuring that it survives the stress of capture and transport. This is generally a much easier task with pinnipeds than with cetaceans. Seals and sea lions are accustomed to being on or near land and tend to adjust readily to a captive environment better than porpoises and whales.

In regard to cetaceans, an application of the knowledge of the natural history of cetaceans is of paramount importance. The size, shape, and depth of pools can mean the difference. Human attentiveness is also important. The primary concern when receiving a captured animal is avoidance of "capture shock", getting the animal to accept the available diet, and attempting to alleviate debilitating injuries, parasitic infestations, and other disease processes which are generally found in the wild. Usually, it will take several days of coaxing and trials before the newly acquired animal will adjust to a collation of deceased edibles, rather than fresh, live, and reluctant fish. Some of the food varieties that are used and seem to be adequate include mackerel, bonito, squid, herring, smelt, and anchovies.[12] Vitamin supplementation is widely practiced and often is empirical in application, rather than based on strict scientific evidence. One investigator has found that certain fish (e.g., herring, smelt) contain thiaminase and thus, certainly, supplementation with thiamine is justified.[11] Porpoises sometimes may take fish into their mouths and then, evidently through some device of evaluation, reject them. This behavior may be a mere nonacceptance of the particular offering, or may indicate disease in newly obtained animals. The continuance of food rejection is obviously a poor prognostic sign, and along with other behavior, such as continual circling of the tank without evident awareness of surroundings, thin or mucoid feces, partially closed eyes, hyperpnea, and lethargy, indicates concern should be immediate. The classic "capture shock" in cetaceans has been compared to capture myopathy as seen in other species. One should routinely sample muscle tissue for histopathology on animals that are necropsied, and with luck, may find areas of hemorrhagic necrosis of muscle tissue (Figure 9).

As indicated, pinnipeds adjust with more ease to captivity. Signs of maladjustment tend to be rather obvious in these animals, and they may be anorectic, lethargic, and may have diarrhea and hyperpnea. Seals and sea lions will thrive on a variety of fish and squid. They are supplemented with vitamins for the same reasons that apply to cetaceans. To our knowledge, "capture shock" has not been seen in pinnipeds, although there has been a report of diffuse muscle degeneration in harbor seals that became clinically manifest approximately two months after capture.[14] No specific cause could be determined. Seals and sea lions that die soon after capture are usually the victims of parasitism, malnutrition, injury, or infectious disease.

Injury associated with capture will usually involve lacerations or abrasions of the skin, and these will generally heal spontaneously if holding tank water is clean (Figure 10). Because of the rather streamlined shape of both pinnipeds and cetaceans, limb injuries are not common, although we have seen fractures of jaw and flipper in California sea lions (Figures 11 and 12). Osteomyelitis has been reported in the flipper of an Atlantic bottlenosed dolphin, subsequent to injury.[15] Cornell has treated several cases of sequestral carpii phalanges in harbor seals, California sea lions, and one killer whale, and a healed fractured coccygeal vertebrae in a dolphin (Figure 13).

V. PARASITES

The first question that must be asked is whether most organisms classified as parasites in marine mammals are, in fact parasites, or are more symbiotic in nature. In the majority of animals in which we identify parasitism, that host animal is going about its business quite nicely, providing a living for one or more smaller creatures that live in or on the host. When watching close, the huge grey whales sound, as they proceed with their annual migration, one is surprised by the quantity of barnacles that reside on the skin of these seemingly well functioning behemoths. The parasite which kills its host cannot be considered, anthropomorphically, a very smart intruder, yet many of them do. Most of us in the field of veterinary medicine have seen apparently normal healthy animals which are parasitized. It has been speculated that there may be immune mechanisms operating in some parasitic infections, resulting in rejection of the parasites as the host matures. It would seem reasonable, for example, that the common lungworm (*P. decorus*) of the California sea lion, would infect almost all animals in the area where the intermediate host (a fish) is found. Yet, while many sea lions develop a terminal pneumonia and are seen gradually achieving entropy on rocks or beaches, no trace of the lungworm can be found in many of the animals which are examined.

In nature, parasites probably operate as one of the myriad of factors which function in choosing the survival of the fittest. In stranded animals, severe parasitism may co-exist with other debilitating diseases, and the major cause of stranding and death may not be easy to sort out. The pathogenicity of a parasite will depend upon a number of factors including the location in the body where the parasite exerts its major influence, the general health and nutritional state of the host, the volume of parasites in or on the host, and any endogenous or exogenous chemical or immune effects that the parasite may present to the host. Most feral marine mammals will probably be found to harbor one or more parasites. Both pinnipeds and cetaceans have been reported to be host to an array of species within the major classes of helminthic parasites (trematodes, cestodes, and nematodes), and pinnipeds can harbor lice, certain mites,[2] coccidia,[16] and sarcosporidia.[17] A harbor seal stranded on a Virginia beach was found to have heartworms (*Dipetalonema spirocauda*), stomach worms (*Contracaecum osculatum*), alopecia associated with pediculosis (*Echinophthirius horridus*), and nasal mites (*Halarachne miroungae*).[18] The authors of this report did not speculate as to the definitive cause of death, but one could assume that the host simply had nothing left to give. To sort out probable cause of mortality, one should possibly arrange parasites found in order of known lethal influence, and then come to a diagnostic consensus, since the specific terminating feature is often obscure in parasitized marine mammals.

In many parasitisms of deceased marine mammals, a specific diagnosis may not be evident since: (1) there are multiple parasitic species involved; (2) the extent of pathogenicity of one or more of the organisms is not well defined; (3) even in some rather severe parasitic infestations the animal may have expired from obscure toxicities, infections, metabolic diseases, or lack of food supply, and; (4) often the most important factor is to determine if the animals' condition is a result of parasites, or is the parasitism a result of the animals condition? Thus, for the pathologist who deals with a patient who was not carefully tended and documented as an inpatient, and whose medical judgement must come after the fact, the contribution of parasites to morbidity and mortality is frequently treated as a matter of relative probabilities.

Once the marine mammals become captive, the speculation as to whether a parasite is pathogenic or not is no longer important to the course which is taken. All exogenous and endogenous organisms that are found to have taken up residence in or on the host are considered to be potential pathogens, and efforts are made to eliminate these intruders.

A variety of medicines have been employed to eliminate parasites, generally without evident detriment to the host. Some examples in the pinnipeds include disophenol against heartworm,[19] thiabendazole, levamisole, and diethylcarbamazine against gastro-intestinal nematodes[20] (diethylcarbamazine may be effective against respiratory and cardiovascular nematodes), niclosamide against intestinal cestodes,[20] dichlorvos against hookworm,[21,22] and disophenol against hookworm and lice.[23] Piperazine is generally effective against nematodes, although toxicity has been reported in two California sea lions.[24]

In cetaceans, thiabendazole, piperazine, and levamisole, as well as Panacur® and Droncit® have been used against gastric nematodes, niclosamide against intestinal cestodes, and lorothidol against gastric, biliary, and nasal trematodes. Both levamisole and lorothidol have produced some side effects in the host, but are quite reliable in killing the target parasite.[12]

Because of the limited number of infected animals available for drug trials, and the continuing development of new parasiticides, the putative course for treatment of parasites will be ever evolving experimentation. This experimentation will probably remain within the domain of the attending veterinarian, since it is unlikely that pharmaceutical manufacturers will ever run toxicity trials on statistically significant numbers of pinnipeds or cetaceans.

VI. INFECTIONS, ANTIBIOTICS, AND VACCINATION

A newly collected animal may be found to have an infectious disease at the time of capture, or shortly thereafter (Figures 14 and 15). Excluding parasitism or "capture shock", these animals may be seen with bacterial, viral, or mycotic conditions. The situation regarding immunity must be considered, in that the spectrum of immunogenic micro-organisms in a pelagic or littoral environment are probably of a different order of magnitude than that on land. Both cetaceans and pinnipeds possess an active reticuloendothelial system similar to their landed, mammalian brethren.[9] Several of the immunoglobulins (IgG, IgM, and IgA) have been identified in both cetaceans and pinnipeds.[25] A study involving a limited number of whales and porpoises indicate that these species have an active cellular immune response.[26] Since marine mammals possess a functioning immune system response, do some of the infections which are seen shortly after capture occur because the animal is exposed to a previously unencountered pathogen? This question can be partially answered in those animals which are found to have stranded as the probable result of a specific disease.

Viral, bacterial, and mycotic diseases have been reported in feral marine mammals. Examples include viral hepatitis in California sea lions,[4] San Miguel sea lion virus (suspected cause of a vesicular skin lesion),[27] dolphin pox,[28] leptospirosis in northern fur seals,[5] salmonellosis in California sea lions,[29] coccidioidomycosis in a sea otter,[1] and aspergillosis in a pygmy sperm whale.[30]

When the animal becomes adjusted to a controlled environment, the cause of morbidity and mortality have appeared most often to involve skin, the respiratory system, septicemia, and to a lesser extent, gastrointestinal problems. There are a multitude of morbid conditions that probably arise from time to time, but unless the animal is unfortunate enough to require necropsy, they pass with only transient episodes of abnormal behavior, such as inappetance or lethargy. It is often difficult at necropsy examination to determine the exact etiology of, for instance, pneumonia, because mixed bacterial cultures are much more common than the isolation of a single organism. One study, based on captive cetaceans, found *Streptococcus, Staphyloccocus, E. coli*, and *Pseudomonas* sp., to be the most commonly isolated bacteria from the lungs of deceased animals.[31] We have also found *Vibrio* sp., *Pasturella*, and *Eromonas* in lungs of marine mammals. Whether these organisms were the definitive cause of pneumonia, act synergistically, or were opportunists remain moot questions.

Erysipelothrix rhusiopathiae may also cause pneumonia and/or septicemia. In our experience, pure cultures of *Erysipelothrix* can best be obtained by culturing spleen or liver (after first searing the area where the sample is to be obtained). Enteritis in porpoises, associated with pasteurella has been reported,[32] and *Clostridium* sp. has been cultured from myositis in marine mammals.[33]

We have seen pinnipeds with intestinal ulcers, where no parasite could be found. Often, these animals were in training, and it has been speculated that stress may be a related factor, causing gastric hyperacidity and ulceration such as occurs in swine, and probably man.

Few virus diseases have been specifically identified in marine mammals, although a number of viruses have been isolated from these species.[34] Seal pox virus,[35] San Miguel sea lion virus,[36] pox virus in cetaceans,[37] an adenovirus hepatitis in sea lions,[4,30] and influenza virus infection in seals,[39] are examples of viral disease so far described. Some examples of mycotic infections which apparently arise after capture are cyclic dermatitis in pinnipeds, associated with *Fusarium* sp.,[10] mucocutaneous candidiasis in cetaceans,[40] cryptococcosis in a porpoise,[41] hepatic nocardiosis in cetaceans,[42] and pulmonary aspergillosis in a porpoise.[43]

Skin disorders are rather common in both pinnipeds and cetaceans in captivity, as well as in the wild, and some have been mentioned previously. Pinnipeds will sometimes develop focal areas of alopecia and this has been associated with friction from rubbing against the sides of a holding pool while swimming a routine pattern (Figure 16). Demodectic mange has also been a cause of dermatitis in pinnipeds.[44] In the case of cetaceans, many of the ulcerative and eruptive skin lesions are of unknown etiology, and present a hodge-podge of bacteria when cultured (Figures 17, 18, and 19). When examined histologically, they appear as a nonspecific dermatitis. Some of these lesions are undoubtedly traumatic in origin, and others may be a result of either impure or osmotically inadequate water. Erysipelas and streptococcosis have been declared primary etiologic causes of some dermal infections.[32] Mycotic dermatitis in a porpoise has been attributed to *Trichophyton* sp. of fungus.[45]

As with domestic species, the therapeutic approach to marine mammals is indeed empirical. Regimens generally are those favored, through experience, by whoever the therapist happens to be. This is not a matter of eccentricity, but rather a reflection of necessity, since reliable drug trials in marine mammals are quite rare. One proceeds on the assumption that "dirty water is bad, antibiotics are good, and breathing is important". A veterinarian who treats marine mammals must be Promethian in approach, and attempt to match therapy vs. disease, in an environment where not only is a diagnosis difficult, but where the patient spends all, or most of its time, in the water (Figure 20). The key here, as in all medicine, is visibility and improvisation, important qualities in a marine mammal clinician (Figure 21). Vaccination against erysipelas and clostridial disease in marine mammals has been recommended (Figure 22).[33,40]

The efficacy of these vaccinations in preventing the conditions has not been proven. However, in clinical trials years ago at Marineland, we found killed bacteria better than modified live erysipelas vaccine to be preferred in dolphins (LC).

VII. CONSIDERATION OF DISEASES CONTAGIOUS TO MAN AND OTHER MAMMALS

Some mention of possible disease causing agents that may be communicable to land mammals has been made, based upon a number of studies. These agents include *Leptospira* sp., *Pseudomonas* sp., *Clostridium* sp., *Pasteurella* sp., and calici-viruses.[46] Others have isolated *Staphylococcus* concurrently from captive dolphins and their human handlers, although cross infection was not proven.[47] Antibodies to *Toxoplasma gondii* have been identified in a California sea lion.[48] Several workers became ill with leptospirosis, apparently contracted from a sea lion which they necropsied, and from which *Leptospira pomona* was isolated.[46] Conjunctivitis in humans has been attributed to influenza type A virus, isolated from seals in which the virus caused pneumonia.[49] Simpson developed a severely enlarged, erythematous, painful thumb, and accompanying fever while handling both pinnipeds and cetaceans. The condition was diagnosed by the attending physician as erysipeloid ("Fish Handlers Disease" and "Blubber Finger"). Though the disease was not confirmed by laboratory tests, it responded well to antibiotic therapy. In humans erysipeloid is caused by *Erysipelothrix rhusiopathiae*, a pathogen for both pinnipeds and cetaceans. Calici-viruses have been isolated from pinnipeds, and it has been speculated that one of these, the San Miguel sea lion virus, may be identical to the virus that causes vesicular exanthema in swine.[27] Also, on at least two occasions, handlers and dolphins had concurrent infections with influenza-like disease, although neither were laboratory diagnosed.

Other than sporadic reports of zoonotic diseases which almost always occur in those persons with "hands on" association with marine mammals (oceanarium personnel, researchers, veterinarians), no significant threat to public health appears to be imminent. Currently, the potential for harm is much greater to the whales, porpoises, seals, and sea lions from man, than from these creatures to humans, and certainly the human population is in greater danger from earthquakes, tornadoes, and exploding volcanoes than from zoonotic disease attributed to any zoological source.

In summary, it is evident that the marine mammal practitioner must be part natural historian, versatile clinician, researcher and inventor to have continued success in marine mammal practice. In addition, he or she must also be aware of new drugs and techniques, and must be a concerned ecologist — a rather awesome and yet enjoyable challenge.

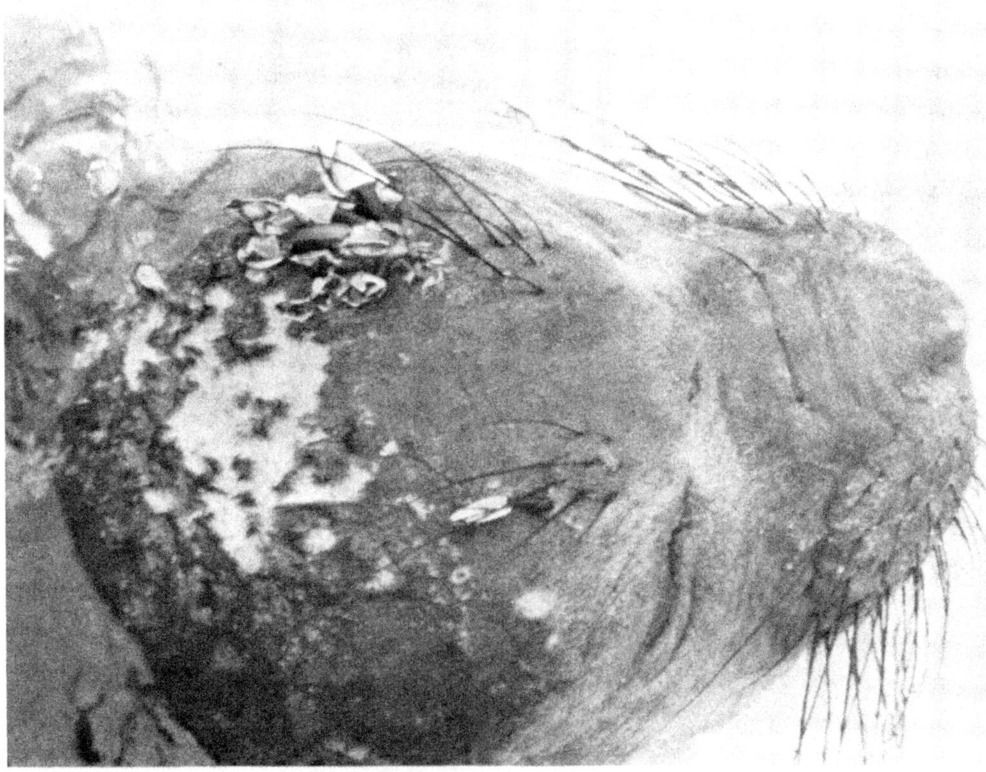

FIGURE 1. A young stranded northern elephant seal showing attached barnacles and skin lesions on the head.

FIGURE 2. This is a young pinniped found dead of gunshot wounds through the head and posterior chest area.

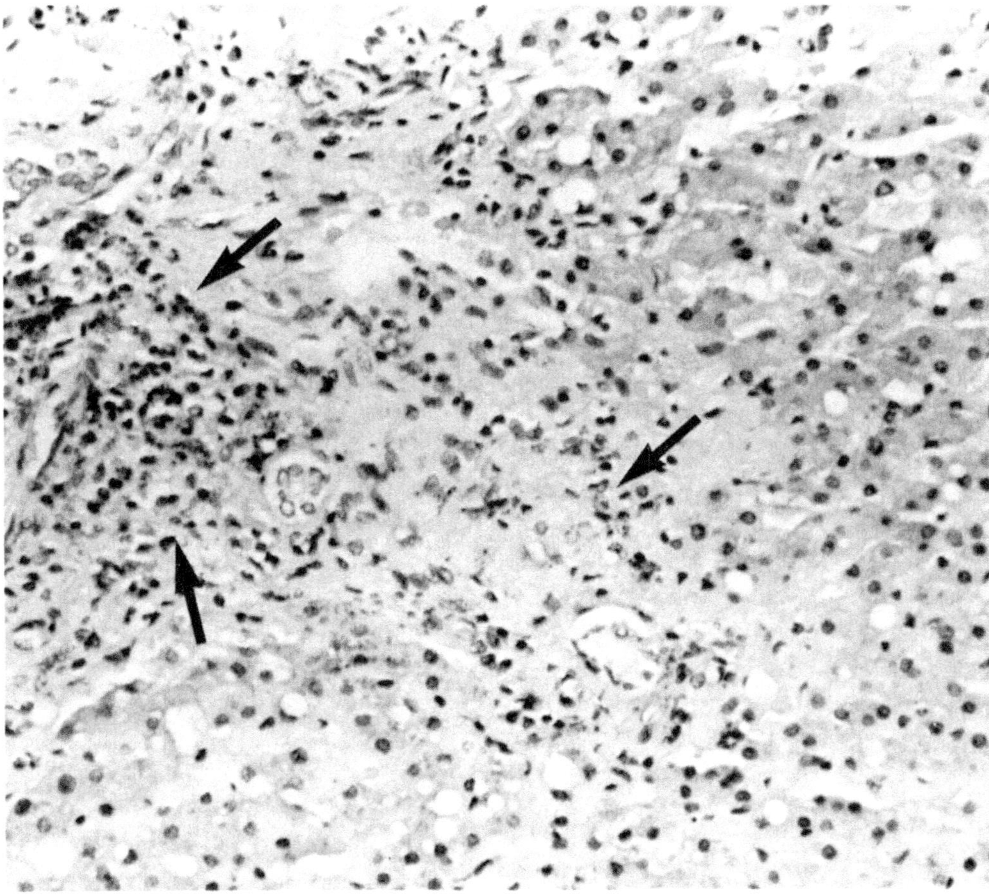

FIGURE 3. Section of liver tissue from a sperm whale, showing focal necrosis and a mixed inflammatory cell infiltrate (arrows).

FIGURE 4. Section of skin from a dolphin, showing ballooning degeneration of the epidermis, caused by being placed in a freshwater tank.

FIGURE 5. Necropsy of an elephant seal, with the opened stomach showing various foreign objects found incidentally during examination.

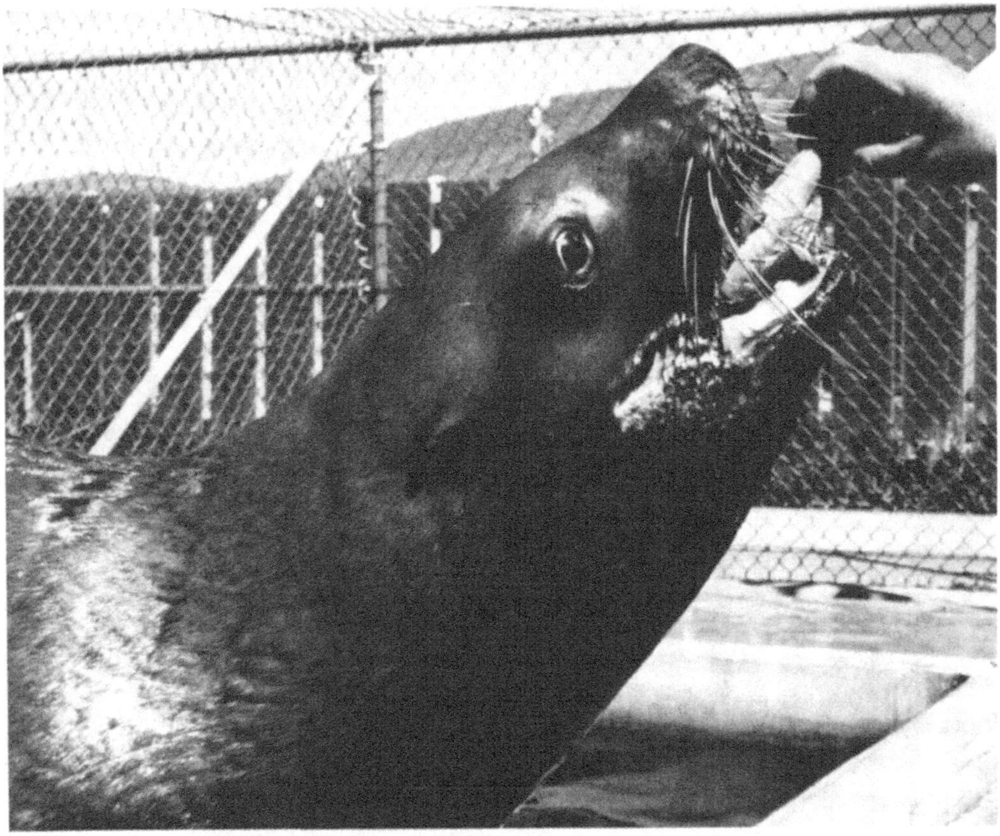

FIGURE 6. Ulceration at the commissure of the mouth of a stellar sea lion which has the appearance of lesions associated with vitamin deficiency in other species. The lesion responded well and was soon gone, after vitamin supplementation was initiated.

FIGURE 7. Attempt to feed a hot dog to a captive dolphin.

FIGURE 8. Feeding fish to a stellar sea lion at an oceanarium in Japan.

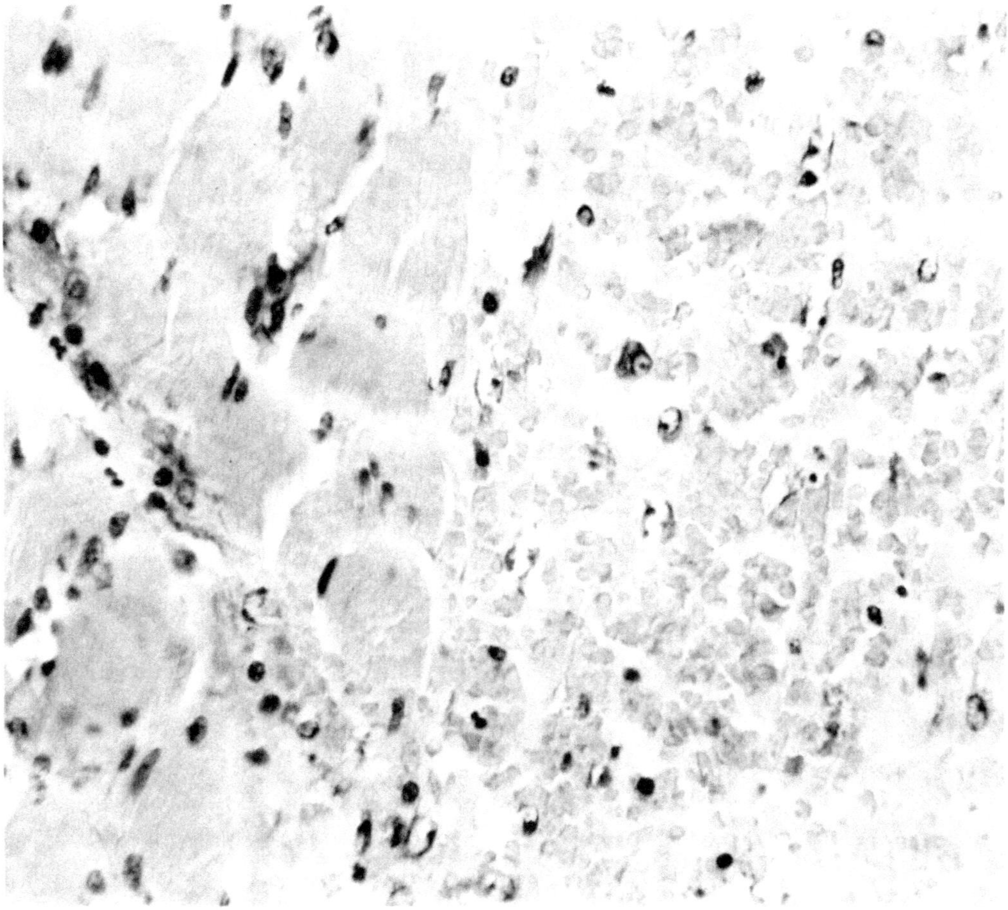

FIGURE 9. Section of skeletal muscle from a dolphin, showing acute hemorrhagic necrosis of muscle tissue, probably representing a typical lesion of capture myopathy.

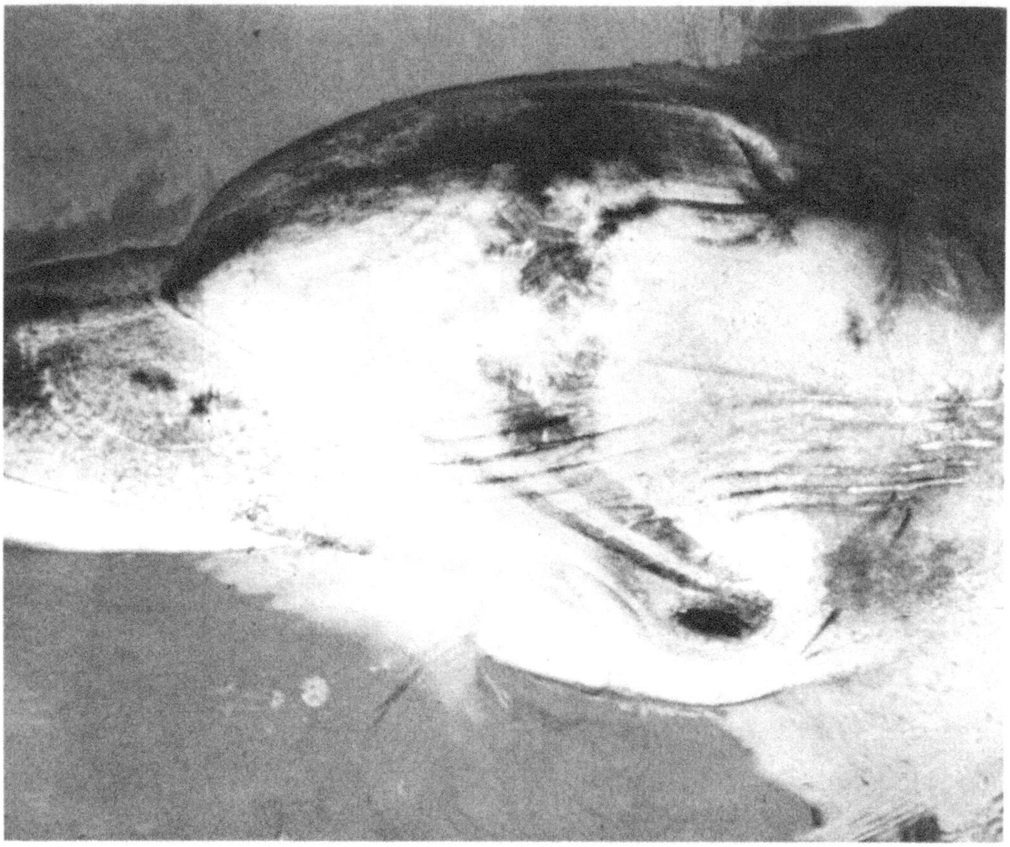

FIGURE 10. Multiple skin lesions of the head of a dolphin, some showing a parallel array of lacerations, suggesting a shark attack.

FIGURE 11A. A young stranded California sea lion with a fractured mandible, probably the result of a traumatic injury. The animal would soon die of starvation in this condition.

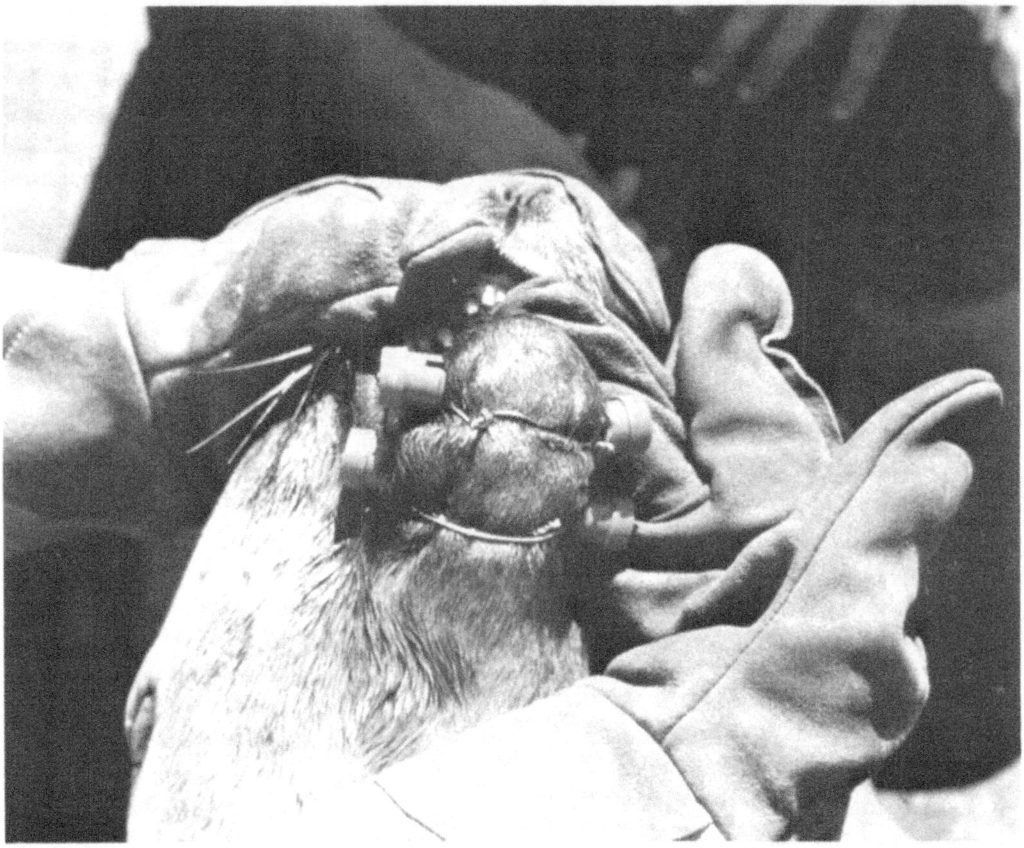

FIGURE 11B. Picture of the same sea lion showing the method of fixation of the fractured mandible. With nursing care and feeding, the animal recovered nicely.

FIGURE 12. This is a dolphin which had suffered a fracture of the upper jaw, with resultant deviation and inability to close the jaws.

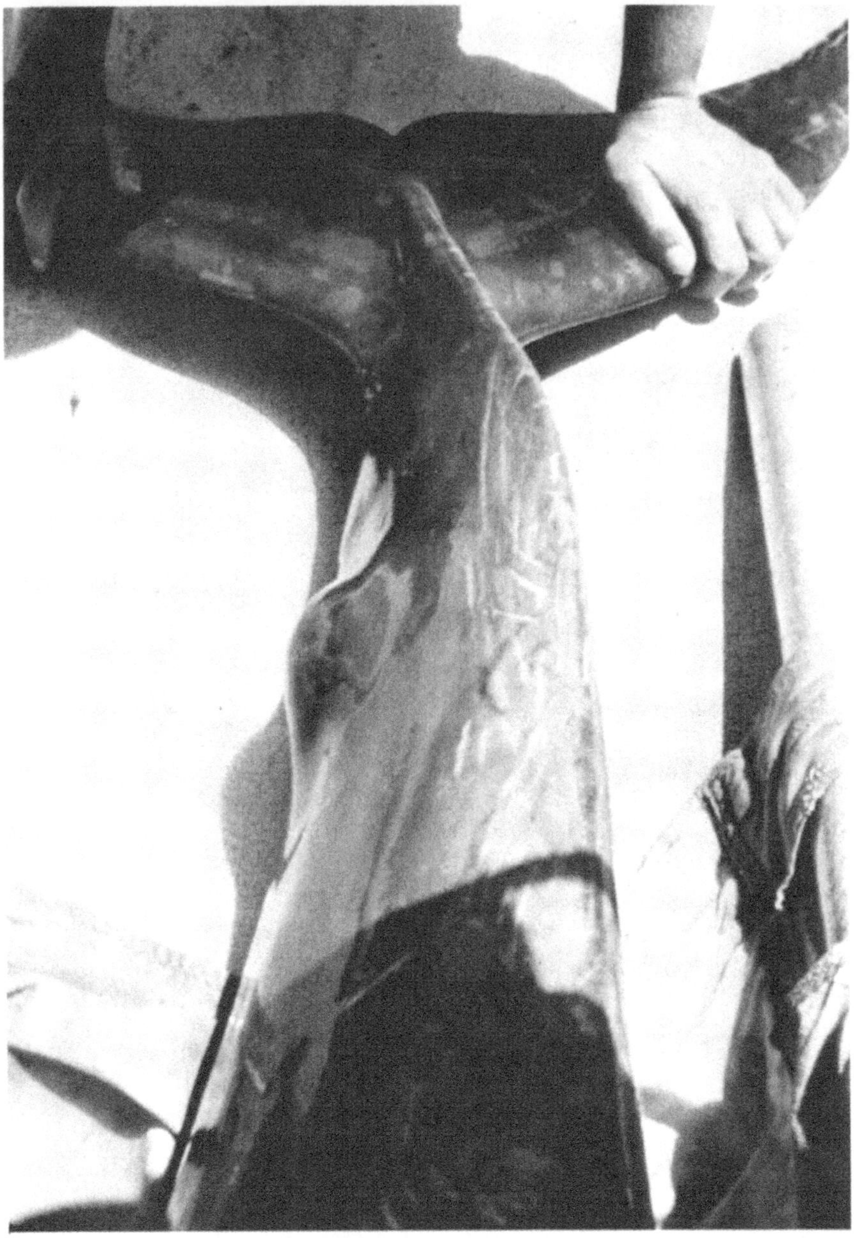

FIGURE 13. Healing fractured coccygeal vertebrae in a dolphin, showing the lateral bulging from the swelling and callous formation.

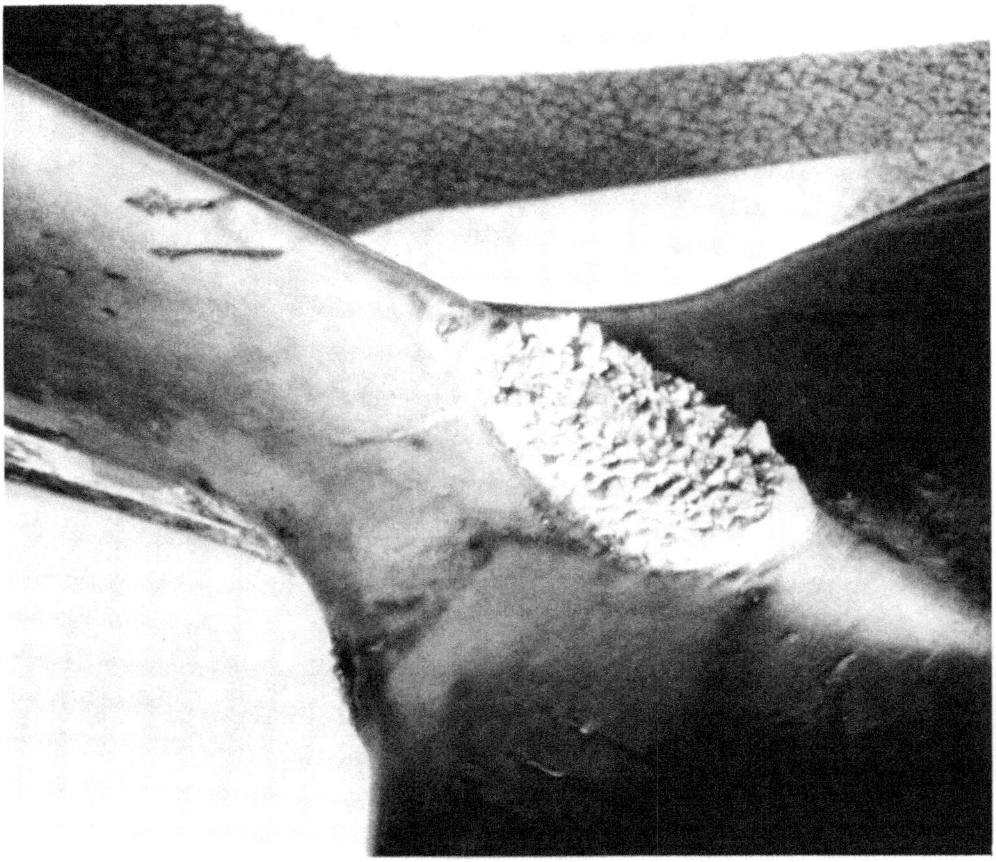

FIGURE 14. This shows a chronic ulcerated and granulating skin wound on the caudal pedicle of an Atlantic bottle-nose dolphin.

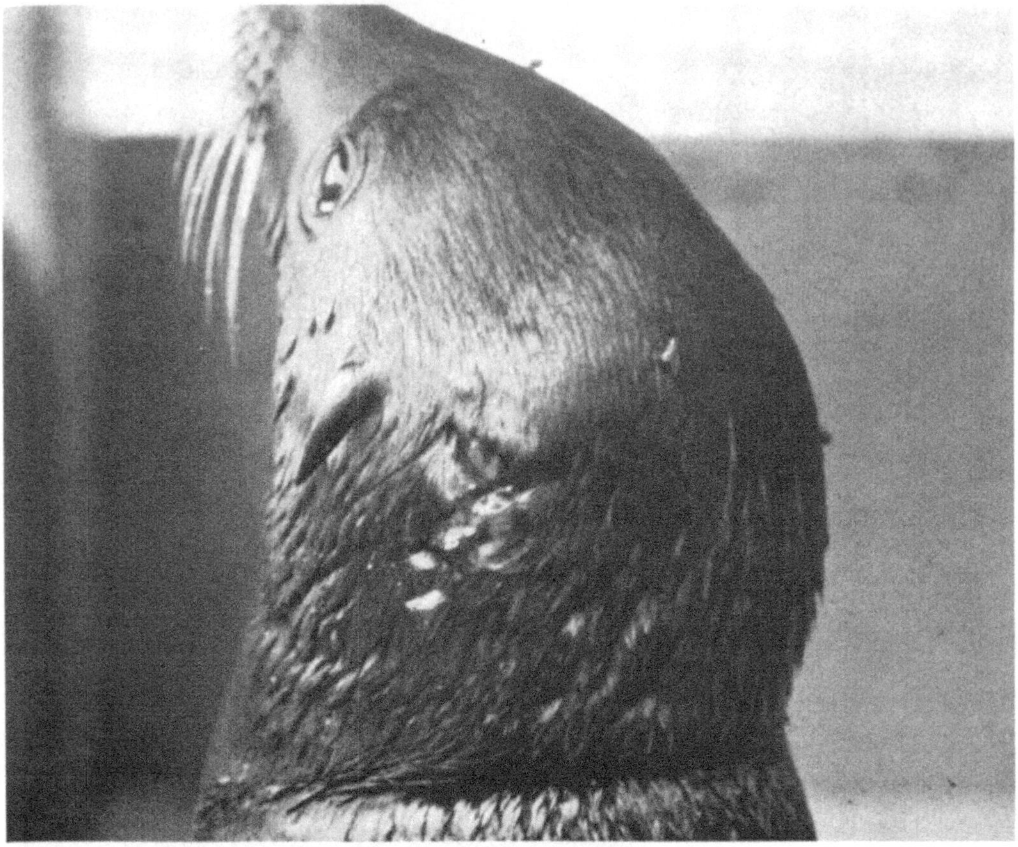

FIGURE 15. Several deep infected lacerations behind the "ear" of this sea lion pup.

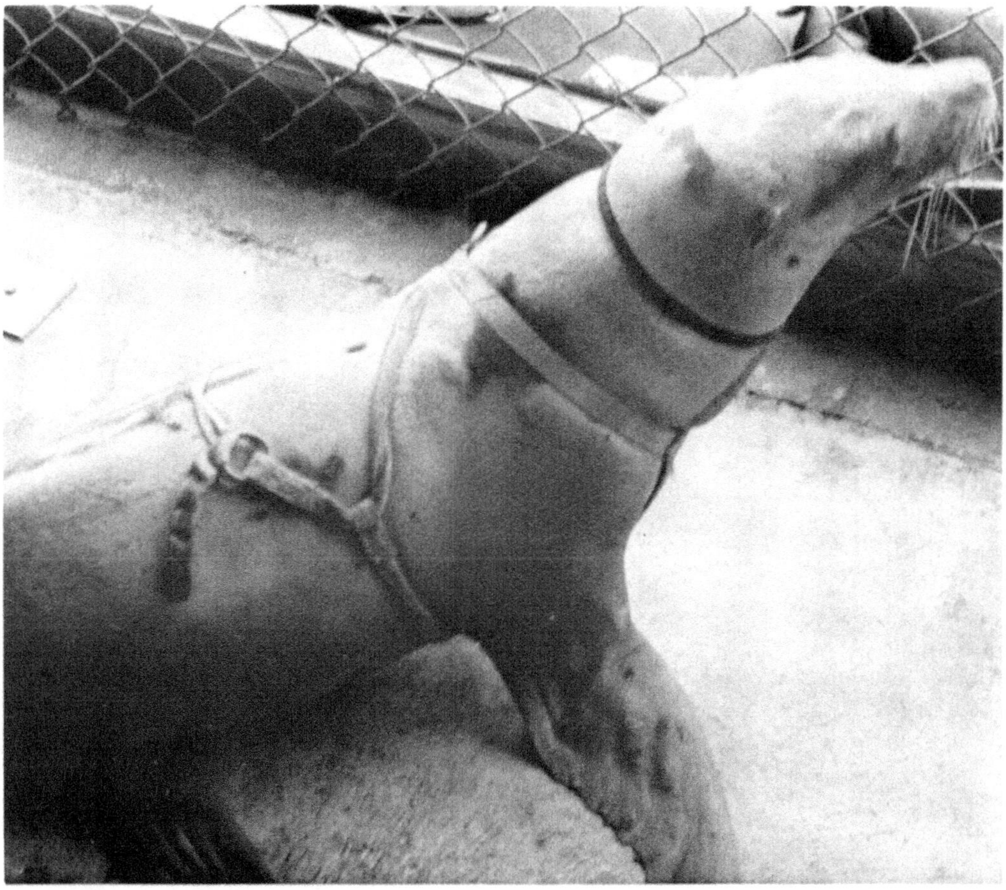

FIGURE 16. Multiple pigmented skin lesions on a sea lion, which represent healed skin erosions from rubbing on the side of the pool. Skin lesions commonly become pigmented on healing in pinnipeds.

FIGURE 17. Multiple focal skin ulcers, of unknown etiology, but bacterially contaminated, on a dolphin.

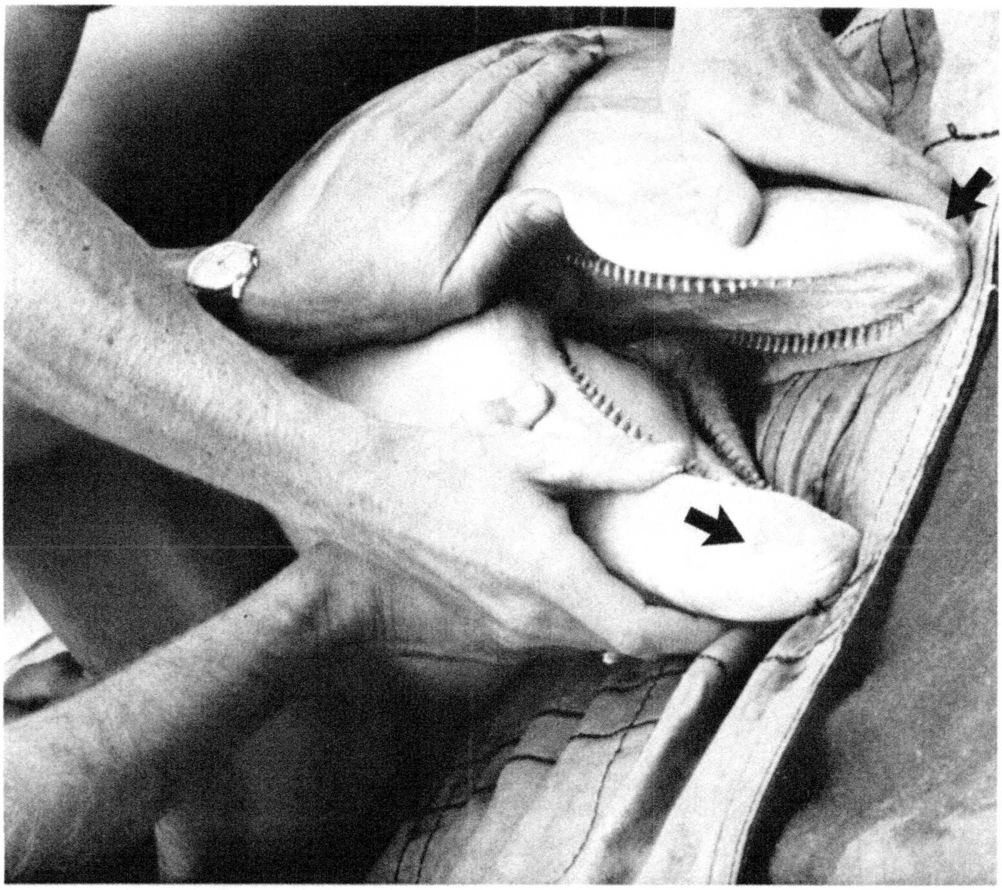

FIGURE 18. Rostral skin erosions in an Atlantic bottle-nose dolphin, attributed to rubbing against the wall of the tank.

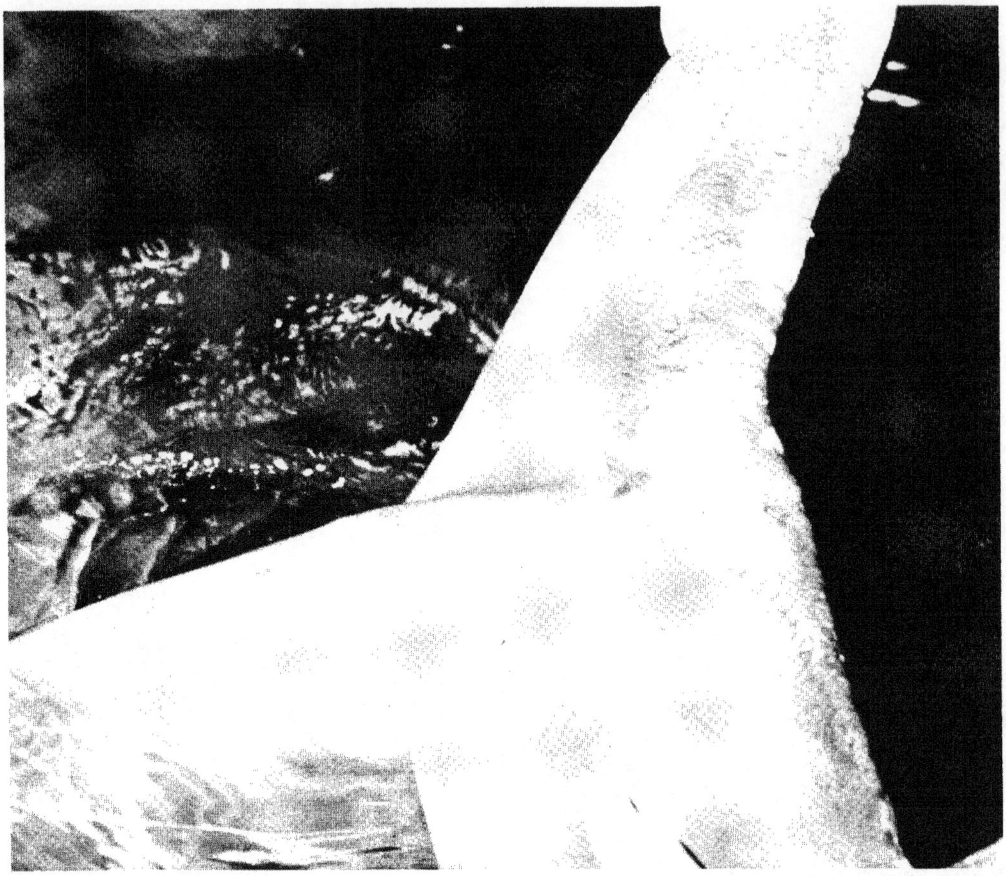

FIGURE 19. Chronic granulating skin erosion, of unknown etiology, on the caudal pedicle of an Atlantic bottlenose dolphin.

FIGURE 20. Killer whale presenting so that a routine blood sample may be taken for laboratory examination.

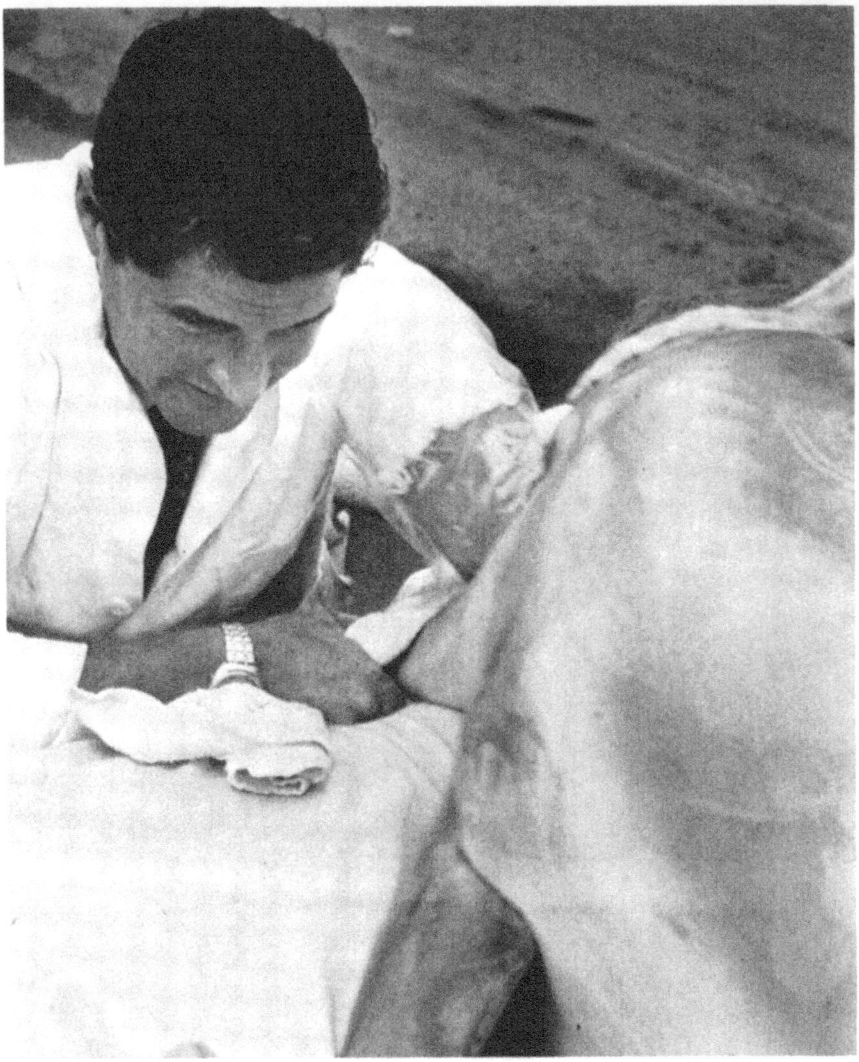

FIGURE 21A. Palpation of the esophagus in a dolphin. This technique is used to examine for foreign bodies, ulcers, lesions, etc.

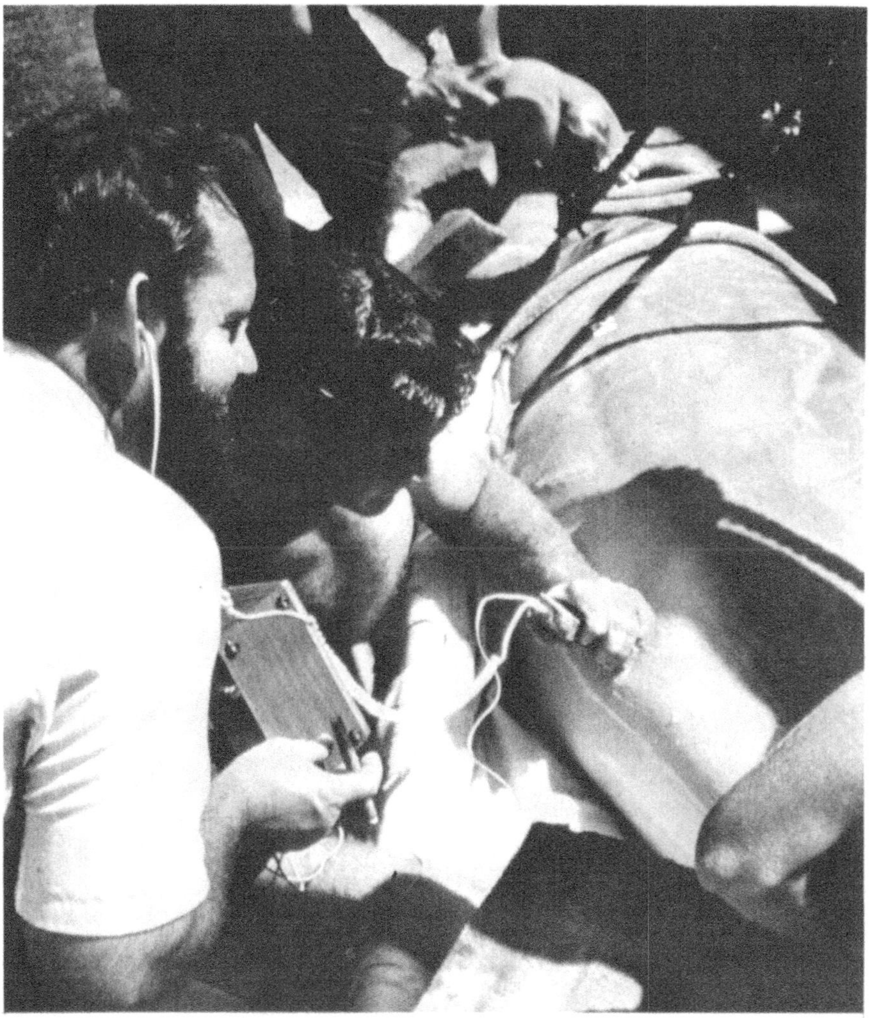

FIGURE 21B. Using a Doppler stethoscope for the detection of a fetal heart beat. This is usually effective only after the 4th month of gestation.

FIGURE 22. A killer whale performing a "slide out" maneuver for routine vaccination procedures.

REFERENCES

1. **Cornell, L. H., Osborne, K. G., Antrim, J. E., and Simpson, J. G.,** Coccidioidomycosis in a California sea otter, *J. Wildlife Dis.*, 15, 373, 1979.
2. **Daily, M. B., and Brownell, R. L., Jr.,** A checklist of marine mammal parasites, in *Mammals of the Sea: Biology and Medicine*, Ridgway, S. H., Ed., Charles C Thomas, Springfield, Ill., 1972, 528.
3. **Geraci, J. R.,** The enigma of marine mammal strandings, *Oceanus*, 21, 38, 1978.
4. **Britt, J. O., Nagy, A. Z., and Howard, E. B.,** Acute viral hepatitis in California sea lions, *J.A.V.M.A.*, 175, 921, 1979.
5. **Smith, A. W., Brown, R. J., Skilling, D. E., Bray, H. L., and Keyes, M. C.,** Naturally occurring leptospirosis in northern fur seals *(Callorhinus ursinus)*, *J. Wildlife Dis.*, 13, 144, 1977.
6. **Stroud, R. K. and Roffe, T. J.,** Causes of death in marine mammals, *J. Wildlife Dis.*, 15, 91, 1979.
7. **Ridgway, S. H. and Dailey, M. B.,** Cerebral and cerebellar involvement of trematode parasite in dolphins and their possible role in stranding, *J. Wildlife Dis.*, 8, 33, 1972.
8. U.S. Dept. of Agriculture Marine Mammal Regulations, *Fed. Register*, 44:36868, 1979.
9. **Simpson, J. G. and Gardner, M. B.,** Comparative microscopic anatomy of selected marine mammals, in *Mammals of the Sea: Biology and Medicine*, Ridgway, S. H., Ed., Charles C Thomas, Springfield, Ill., 1972, 298.
10. **Montali, R. J., Bush, M., Strandberg, J. D., Janssen, D. L., Boness, D. J., and Whitla, J. C.,** Cyclic dermatitis associated with *Fusarium* sp. infection in pinnipeds, *J.A.V.M.A.*, 179, 1198, 1981.
11. **Geraci, J. R.,** Dietary disorders in marine mammals: synthesis and new findings, *J.A.V.M.A.*, 179, 1183, 1981.
12. **Sweeney, J. C. and Ridgway, S. H.,** Procedures for the clinical management of small cetaceans, *J.A.V.M.A.*, 167, 540, 1975.
13. **Colgrove, G. S.,** Suspected transportation — associated myopathy in a dolphin, *J.A.V.M.A.*, 173, 1121, 1978.
14. **Wilson, T. M.,** Diffuse muscular degeneration in captive harbor seals, *J.A.V.M.A.*, 161, 608, 1972.
15. **Foley, R. H.,** Osteomyelitis of the flipper of a bottle-nosed dolphin, *J.A.V.M.A.*, 175, 999, 1979.
16. **Chao-Kuang Hsu,** Coccidiosis in the harbor seal, *J.A.V.M.A.*, 163, 1207, 1973.
17. **Migaki, G. and Albert, T. F.,** Sarcosporidiosis in the ringed seal, *J.A.V.M.A.*, 177, 917, 1980.
18. **Conlogue, G. J., Ogden, J. A., and Foreyt, W. J.,** Pediculosis and severe heartworm infection in a harbor seal, *Vet. Med.*, 75, 1184, 1980.
19. **Brown, R. J., Smith, A. W., Keyes, M. C., Trevethan, W. P., and Kupper, J. L.,** Lesions associated with fatal hookworm infection in the northern fur seal, *J.A.V.M.A.*, 165, 804, 1974.
20. **Sweeney, J. C.,** Procedures for clinical management of pinnipeds, *J.A.V.M.A.*, 165, 811, 1974.
21. **Bigg, M. A. and Lyons, E. T.,** Clinical observations on three northern fur seal pups treated with dichlorvos, *J.A.V.M.A.*, 179, 1284, 1981.
22. **Lyons, E. T., Keyes, M. C., and Conlogue, J.,** Activities of Dichlorvos or disophenol against the hookworm (*Uncinaria lucasi*) and sucking lice of northern fur seal pups (*Callorhinus ursinus*) on St. Paul Islands, Alaska, *J. Wildlife Dis.*, 14, 445, 1978.
23. **Lyon, E. T., Kim, K. C., and Keyes, M. C.,** Variable activity of disophenol against hookworms and lice on northern fur seal pups on St. Paul Islands, Alaska, *J. Wildlife Dis.*, 16, 53, 1980.
24. **Gray, K. N.,** Piperazine toxicity in California sea lions *(Z. Californianus)*, *J. Zoo An. Med.*, 3, 44, 1972.
25. **Cavagnolo, R. Z.,** The immunology of marine mammals, *Devel. Comp. Immunol.*, 3, 245, 1979.
26. **Mumford, D. M., Stockman, G. D., Barsakes, P. B., Whitman, T., and Wilbur, J. R.,** Lymphocyte transformation studies of sea mammal blood, *Experientia*, 31, 498, 1975.
27. **Sawyer, J. C.,** Vesicular exanthema of swine and the San Miguel sea lion virus, *J.A.V.M.A.*, 169, 707, 1976.
28. **Geraci, J. R., Hicks, B. D., and Aubin, D. J.,** Dolphin pox: a skin disease of cetaceans, *Can. J. Comp. Med.*, 43, 399, 1979.
29. **Schroeder, R. J., Delli Quadri, C. A., McIntyre, R. W., and Walker, W. A.,** Marine mammal disease surveillance program in Los Angeles County, *J.A.V.M.A.*, 163, 580, 1973.
30. **Sweeney, J. C., Migaki, G., Vainik, P. M., and Conklin, R. H.,** Systemic mycoses in marine mammals, *J.A.V.M.A.*, 169, 946, 1976.
31. **Medway, W. and Schryver, H. F.,** Respiratory problems in captive small cetaceans, *J.A.V.M.A.*, 163, 571, 1973.
32. **Sweeney, J. C. and Ridgway, S. H.,** Common diseases of small cetaceans, *J.A.V.M.A.*, 167, 533, 1975.
33. **Greenwood, A. G. and Taylor, D. C.,** Clostridial myositis in marine mammals, *Vet. Rec.*, 103, 54, 1978.
34. **Smith, A. W. and Skilling, D. E.,** Viruses and virus diseases of marine mammals, *J.A.V.M.A.*, 175, 918, 1979.
35. **Wilson, T. M., Dykes, R. W., and Tsai, K. S.,** Pox in young captive harbor seals, *J.A.V.M.A.*, 161, 611, 1972.

36. **Smith, A. W. and Akers, T. G.,** San Miguel sea lion virus isolation, preliminary characterization and relationship to vesicular exanthema of swine virus, *Nature (London)*, 244, 108, 1973.
37. **Flom, J. O. and Houk, E. J.,** Morphological evidence of pox virus in "tatoo" lesions from captive bottle-nosed dolphins, *J. Wildlife Dis.*, 15, 593, 1979.
38. **Dierauf, L. A., Lowenstine, L. J., and Jerome, C.,** Viral hepatitis (adenovirus) in a California sea lion, *J.A.V.M.A.*, 179, 1194, 1981.
39. **Geraci, J. R., St. Aubin, D. J., Barker, I. K., Webster, R. G., Hinshaw, V. S., Bean, W. J., Ruhnke, H. L., Prescott, J. H., Early, G., Baker, A. S., Madoff, S., and Schooly, R. T.,** Mass mortality of harbor seals: pneumonia associated with influenza A virus, *Science*, 215, 1129, 1982.
40. **Medway, W.,** Some bacterial and mycotic diseases of marine mammals, *J.A.V.M.A.*, 177, 831, 1980.
41. **Migaki, G., Gunnels, R. D., and Casey, H. W.,** Pulmonary cryptococcosis in an Atlantic bottle-nosed dolphin *(Tursiops Truncatus), Lab. Anim. Sci.*, 28, 603, 1978.
42. **Pier, A. C., Takayma, A. K., and Miyahara, A. Y.,** Cetacean nocardiosis, *J. Wildlife Dis.*, 6, 112, 1970.
43. **Carroll, J. M., Jasmin, A. M., and Baucom, J. N.,** Pulmonary aspergillosis of the bottle-nosed dolphin *(Tursiops Truncatus), Vet. Clin. Pathol.*, 2, 139, 1968.
44. **Sweeney, J. C.,** Common diseases of pinnipeds, *J.A.V.M.A.*, 165, 805, 1974.
45. **Hoshima, T., and Sigiura, Y.,** On a skin disease and a nematode parasite of a dolphin, *Tursiops truncatus, Sci. Rep. Whales Res. Inst.*, 1966.
46. **Smith, A. W., Vedros, N. A., Akers, T. G., and Gilmartin, W. G.,** Hazards of disease transfer from marine to land mammals: review and recent findings, *J.A.V.M.A.*, 173, 1131, 1978.
47. **Streitfeld, M. M., and Chapman, C. G.,** Staphylococcus aureus infections of captive dolphins (*Tursiops Truncatus*) and oceanarium personnel, *Am. J. Vet. Res.*, 37, 303, 1976.
48. **Riemann, H. P., Behymer, D. E., Fowler, M. E., Schulz, T., Lock, A., Orthoefer, J. G., Silverman, S., and Franti, C. E.,** Prevalence of antibodies to toxoplasma gondii in captive exotic mammals, *J.A.V.M.A.*, 165, 798, 1974.
49. **Webster, R. G., Geraci, J. R., Petursson, G. and Skirnisson, K.,** Conjunctivitis in human beings caused by influenza A virus of seals, *N. Engl. J. Med.*, 304, 911, 1981.

Chapter 3

THE HEMATOPOIETIC SYSTEM

James O. Britt, Jr. and Edwin B. Howard

TABLE OF CONTENTS

I. INTRODUCTION

The immune system of marine mammals has not been thoroughly studied. Isolated observations and experiments have been performed and reported, but in most instances the animals studied were isolated cases without long-term follow-up available. The levels of pollutants probably varies considerably between coastal areas, and the effects of these chemicals on the immune response may be significant (see Chapter 6, Volume II), but it is ambiguous at this time and undoubtedly there is variation in species and in individual animal responses. Also, it may be erroneous to compare data from different locations, even within the same species, because subpopulations may be affected by variable levels of pollutants in the water, and environmental factors and diseases have a marked effect on the immune system. The external lymph nodes of pinnipeds, especially the scapular and thoracic lymph nodes, are large and prominent between the fascial planes of the musculature. They are pale on the external and cut surfaces, are very moist, and usually 5 to 8 cm in length in subadult sea lions. The tracheobronchial lymph nodes are similarly prominent, although not as large, and numerous mesenteric and sublumbar nodes can be identified in their respective chains. The oropharyngeal tonsils of pinnipeds consist of a relatively small quantity of hypoplastic lymphoid tissue within shallow crypts of the mucosa (Figure 1).

The lymph nodes of pinnipeds have the usual mammalian microscopic anatomy except for hypoplasia of the lymphoid follicles.[1] The nodes of young and subadult pinnipeds consist of a nearly uniform cortical sheet of small lymphocytes, which in some individual animals is vaguely subdivided into primary follicles. Secondary follicles with germinal centers are very rarely seen in our local pinnipeds, even in the nodes draining areas of bacterial infection. The afferent lymphatic vessels appear to drain into the nodes via subcapsular sinuses, as in most mammals. The medullary sinuses of pinniped nodes are prominent, and filled with histiocytes and some plasma cells, with an overall edematous appearance. Lymph nodes that are judged to be reactive due to an increase in size, and are in proximity to an area of inflammation, have an increased population of histiocytes in the medulla and interfollicular areas of the cortex, and the primary lymphoid follicles are larger in size than expected for the inactive state. Reactive pinniped nodes appear especially edematous on gross examination. The gastric lymph nodes for example, in pinnipeds may increase in length from less than 1 cm to 4 to 5 cm when active parasitic gastric ulceration is present, due to lymph node edema and histiocytosis.

Porpoises in general also tend to lack well defined secondary germinal follicles in their lymph nodes, and the cortico-medullary junction is poorly defined microscopically because there are numerous histiocytes in the medulla and in the interfollicular cortex, and the primary follicles are smaller than in pinnipeds. Eosinophils are numerous in the medullary sinuses and interfollicular areas of small cetaceans, which is consistent with the high percentage of circulating eosinophils in the blood of many porpoises. Although it has been postulated that the eosinophilia is a response to parasitism,[1] we have seen pronounced lymph node eosinophilia in porpoises with few or no identified parasites at the time of death, and few eosinophils may be seen in the nodes of parasitized pinnipeds. However, it is possible that the eosinophils are reacting to a past transitory parasitism in these cetaceans, and their immunological response may vary from that of pinnipeds with similar types of parasitic diseases.

The lymph nodes of adult northern fur seals from the Pribilof Islands have been described as having numerous germinal centers in the follicles.[2] These germinal centers, indicating secondary follicles, developed in most fur seals in the tonsils and mesenteric lymph nodes by four months of age, and somewhat later in the spleen. It was postulated that these anatomical differences were due to earlier antigenic stimulation of the lymphoid tissue of the mouth and intestinal tract, compared to the visceral sites and spleen. It is difficult to explain the vigorous germinal follicle development in these fur seals while the marine mammals of southern California coastal waters have underdeveloped nodes. The differences could be due to the fact that we have different species to examine (sea lions, harbor seals, and elephant seals), or due to probably much higher exposure to pollutants of southern California littoral mammals compared to the pelagic fur seals of the remote Bering Sea area.

The splenic lymphoid follicles are also generally devoid of germinal centers in porpoises and pinnipeds, and the white pulp is not plentiful, compared to the red pulp. Lymphoid tissue referred to as ''tonsils'' was mentioned as being present in a few of the normal sperm whales examined.[3] These tonsils were found at the base of the tongue and around the anal opening.

Notwithstanding the relative hypoplastic appearance of marine mammal lymphoid tissue, the nodes of pinnipeds can become markedly reactive. Young fur seals with fatal hookworm infection were reported to have mesenteric nodes that were enlarged 3 to 4 times.[4] These nodes consisted of solid sheets of lymphocytes and histiocytes. Germinal centers were not present, and many Reed-Sternberg like binucleated giant cells were seen (Figure 2). It was suggested that this reaction was a hypersensitivity response to the intestinal parasites, but eosinophils were not seen in these nodes. Infrequently the medullary sinuses of nodes draining areas of chronic infection may be packed with plasma cells (Figure 3). Sea lion lymph nodes react very vigorously in response to metastatic carcinoma cells. Of the cases of bile duct and uterine adenocarcinoma we have examined, there was widespread lymphadenopathy, up to 7 to 8 cm of all intra-abdominal lymph nodes, including the hepatic, mesenteric, gastric, and sublumbar nodes. This widespread lymphadenopathy can often make it impossible to determine which nodes were draining the primary site of neoplasia, after what had probably been a long-term clinic course and dissemination. One sea lion with uterine adenocarcinoma had the usual intra-abdominal lymphadenopathy but the scapular nodes were also enlarged and contained metastases. It appears that the nodes of sea lions are especially effective at trapping circulating malignant cells, or that carcinoma cells of this species have an especially malignant nature or affinity for lymphoid tissue.

Microscopically, these metastases to the nodes resulted in effacement of the follicles with depletion of the population of lymphocytes, and there was necrosis and fibrosis in the cortex and medulla. Microcystic subcapsular sinuses often were seen to be filled with tumor cells, suggesting that the afferent lymphatics empty into these spaces as in most other mammals. Splenic metastases were common and also resulted in destruction of the normal architecture of both the red and white pulp, with necrosis and fibrous replacement.

The lymphocyte subpopulations of all species of marine mammals have not been accurately measured for the relative number of T and B lymphocytes. Mumford stimulated whale lymphocytes with mitogens *in vitro* and found much greater response to pokeweed mitogen than to phytohemagglutinin, suggesting that the whale had more B cells than T cells, assuming these mitogens induce responses similar to those in humans and mice.[5] The humoral immune system has been studied to a limited extent in marine mammals. The IgG and IgM subclasses of immunoglobulins have been identified in sea lions and porpoises.[6,7] This was done using antihuman immunoglobulin antisera, with assumed cross-reactivity to the corresponding animal immunoglobulin classes in precipitation and agar cell diffusion tests. In addition to sharing similar antigenic sites, the marine mammal immunoglobulins examined were similar in size to the human counterparts.

Levels of immunoglobulins have only been quantitated in one species of marine mammal, the northern fur seal.[2] Immunoglobulins G, M, and A were identified and purified by gel filtration chromatography. The IgG had a molecular weight of 150,000 daltons, and a sedimentation coefficient of 7s, identical to human IgG. Immunoglobulins M and A were also found to be similar between human and fur seal; IgM was 1,100,000 daltons (18s) and IgA was 320,000 daltons (12s). The IgA in the serum and in throat mucus was present as a dimer, and serum IgM was a pentamer, similar to other mammalian species. Immunoglobulin E has not been found in any marine mammal, but this may be due to its relatively low level and difficulty in purification. The level of IgE might be postulated to be high compared to terrestrial mammals due to the high degree of parasitism in sea mammals. Perhaps the elusive IgE is important in the phenomenon by which the neonatally acquired hookworm infections of fur seals are spontaneously eliminated at 2 to 3 months of age.[9]

Transfer of maternal antibodies appears to be minimal to the fur seal fetus, according to the recent studies by Cavagnolo.[10] The full term fetus has a serum IgG level of only 5% of the dam and IgM is too large to cross the placenta, but it is present in the newborn pup at about 19% of the average adult level. Immunoglobulins G and M are only moderately concentrated in the dam's colostrum but IgA is concentrated nearly 23-fold. The colostral IgG is very poorly absorbed by the suckling pup, as in other species, and IgM and IgA do not increase at all in colostral fed pups. Serum IgM is produced by the neonate, and the level begins to increase rapidly at one week of age to about 70% of the adult level by five weeks of age. The immunoglobulins may be poorly absorbed from the intestine because the colostrum has a very high fat level (50% v/v), and the lipid molecules may interfere with absorption of the immunoglobulin macromolecules.[10]

Although adult fur seals have immunoglobulin levels similar to other mammals, their pups are very hypogammaglobulinemic until they start producing antibodies after one week of age. It is interesting that no epidemic infectious diseases have been identified in these pups from the crowded rookeries on the Pribiloff Islands. Perhaps the environment is hostile to pathogens, but this would seem unlikely to fully negate the hazards associated with concentration of excrement and enteric bacteria. Possibly serum complement and neutrophil phagocytosis may be relatively more important in these animals, but these parameters have not been investigated.

Hematopoietic neoplasms appear to be the most common malignancy of pinnipeds, based on a review of case reports in the literature.[11] Seven of the cases were of multicentric malignant lymphoma with visceral involvement. One interesting report of malignant lymphoma in two harbor seals suggested the possibility of an infectious etiology since the two animals died within a short period of time, were housed together, and were only $1^1/_2$ and $2^1/_2$-years-old.[12] Both were leukemic, but the neoplasm could not be transmitted to healthy harbor seals by injection of peripheral blood leukocytes or macerated splenic or nodal tissue. A case of apparent malignant lymphoma in an infant fur seal was described as having possible retroviruses budding from cell membranes,[13] but this may have been the hyperplasia and lymphadenopathy associated with hookworm infection. A single case of malignant lymphoma has also been reported in a cetacean, a *Tursiops truncatus* female, older than 20 years, involving the spleen.[11]

We have seen one case of what has been called status thymico-lymphaticus[14] in an adult female sea lion that died suddenly and unexpectedly. She had been in captivity for 18 months after stranding. It is a sudden death syndrome and is associated with primary adrenal cortical atrophy as well as thymic hyperplasia. The thymus is thought to undergo regeneration and hyperplasia (or perhaps it never regressed normally if there had been chronic borderline adrenal insufficiency). In our case, the thymus consisted of a prominent mass of otherwise normal tissue in the anterior mediastinum with V-shaped posterior extensions over the pericardial sac. The adrenal glands were reduced to about 1/2 the normal size, but there was no evidence of inflammation or fibrosis. The heart seemed to be somewhat small, as with Addison's disease of the dog, and the lymph nodes were somewhat hyperplastic and some follicles had germinal centers. Perhaps there had been stimulation of the B lymphocytes by the hyperplastic T lymphocytes. The persistence of a large thymus might be explained by a decreased level of adrenocorticosteroids, which normally tend to cause lymphopenia by T-lymphocytolysis.[15]

II. HEMATOPOIESIS

The main sites of hematopoiesis in marine mammals appear to be in the vertebral bodies. Due to their streamlined anatomy, the long bones of terrestrial mammals are extremely shortened in pinnipeds and cetaceans, so that the quantity of marrow is also reduced. Additionally, the cortical bone is thick and dense, impinging laterally on the potential space available for marrow. The marrow of the vertebral bodies is plentiful and is enmeshed within a spongy network of bony spicules, and marrow fat is not plentiful. Hematopoiesis is also not plentiful in the ribs of the pinnipeds and small cetacea that we have explained, as the medullary cavity is small.

The spleen is often a site of hematopoiesis in pinnipeds and small cetaceans (Figure 4). We suspect that this is a normal physiological mechanism in these animals, rather than a response to a pathological condition (anemia). Presumably this extramedullary site of red cell production is necessary to these animals that have a high hematocrit and blood volume. The red pulp of pinnipeds and small odontocetes is usually quite vascular and congested microscopically, and is usually quite cellular with mixed leukocytes, including histiocytes. Extramedullary hematopoiesis is most common in young and yearling pinnipeds, but is not unusual in older pinnipeds also. In our experience, the spleen of pinnipeds, especially elephant seals grossly have surface fissures that may enable the organ to expand and contract as necessary for blood pooling and release. Otherwise, the spleen of pinnipeds has the usual flattened, elongated shape of other mammals.

The spleen of small odontocetes is round and varies in color from greyish-brown to dark red and purple in various species. It is small and round, about 6 cm in diameter and weighs 20 to 60 g in *Delphinus*,[16] and it is loosely attached to the serosal surface of the stomach near the groove between the first and second chambers. Arvy reviewed the literature and reported that the spleen of *Physeter* and *Grampus* is lobulated, and *G. griseus* may have as many as seven accessory spleens.[16] This information might be helpful in these species, in preventing an erroneous diagnosis of prior splenic injury. Arvy concluded that the most common pattern in small cetacea was a single round spleen, present in about 80% of the *Delphinus* and *Stenella* (Figure 5). The microscopic anatomy of the cetacean spleen is typically mammalian, although germinal centers are infrequent in the lymphoid follicles. These primary follicles are larger in odontocetes than in mysticetes,[16] and may coalesce, resulting in near obliteration of the red pulp. Perhaps the lack of germinal centers is due to lack of adequate antigenic stimulation in these pelagic species. Although the red pulp is very congested, the spleen probably does not serve as a significant site for blood pooling and storage since there are no surface fissure lines, and the spherical shape is not as efficient for expansion as is the elongated, flat pinniped spleen. Also, a large spleen might be detrimental due to susceptibility to rupturing in these deep diving animals. Blood pooling may be more important in the retia mirabilia and in the central venous system in cetaceans.

The kinetics of iron metabolism and hematopoiesis have not been extensively studied in marine mammals.[17] We observed stored tissue iron in the form of Perl's iron positive hemosiderin in the majority of the livers and spleens of elephant seals, and less commonly in other species. The iron granules were seen in the cytoplasm of hepatocytes and Kupffer cells in the liver, and within splenic hemosiderophages in large granules and aggregates (see Chapter 2, Volume I). This hemosiderosis was not associated with any identifiable pathological process such as hemolysis and degenerative liver disease. The hemosiderin may serve as a storage pool of iron in animals that have a high requirement for red blood volume, such as elephant seals with a hematocrit of 60% or more.

A few observations on iron metabolism have been made in small odontocetes. Their total iron binding capacity is high, and was reported for *T. truncatus* as 486 ± 144 μg/100 mℓ,[17] and 516 ± 32 μg/100 mℓ,[18] compared to 250 to 400 μg/100 mℓ in man. Harbor porpoises with a blood picture of hemorrhagic (blood loss) anemia have responded clinically with iron therapy.[19] The porpoises were thought to have suffered from chronic blood loss due to *Pseudalius inflexus* nematodes in the pulmonary vasculature and/or liver flukes, *Campula oblonga*. Newly captured animals have a higher hematocrit, hemoglobin concentration, and serum iron level than they do after five months in captivity.[20] The cause of this change is unknown, but certain fish fed raw to young mink produce microcytic, hypochromic (iron deficiency), anemia which is preventable by boiling the fish, but not by supplementing the diet with vitamins and iron.[18] Perhaps these fish contain something that prevents iron absorption, and a similar problem might be possible in marine mammals being fed a diet containing a limited number of fish species. Also wild porpoises have a higher RBC count because of higher demands for oxygen in the wild.[21] The life span of circulating erythrocytes has not been calculated in any species of marine mammal, to our knowledge.

As can be seen in the following charts of normal blood values (Table 1), the normal pinniped hemogram has very few eosinophils, in contrast to some of the small cetaceans that may have a relative eosinophil count of 30% or more. The granules of pinniped eosinophils are small and not numerous, making their identification somewhat difficult in formalin fixed tissue sections. Eosinophils have been reported to comprise about 10% of the inflammatory cells in the lungs of sea lions with *Parafilaroides decorus* induced pneumonia, with the majority of the leukocytes being histiocytes and neutrophils.[22] Parasitized, stranded sea lions usually do not have a peripheral blood eosinophilia. However, if the eosinophil count is increased to about 10% this is highly suggestive of parasitic pneumonia or ulcerative parasitic gastritis.[23] The white blood cell count is often in the normal range in sea lions with terminal verminous pneumonia, but the count may be elevated to 20,000 to 24,000, and occasionally as high as 60,000. It is difficult to determine if a severe inflammatory reaction seen microscopically against lungworms is complicated by a secondary bacterial pneumonia, because the cellular inflammatory picture is similar whether or not significant bacteria are isolated.

Primary bacterial infections in pinnipeds associated with peritonitis, pneumonia, mastitis, abscesses and septicemia (*Salmonella* or *Edwardsiella*) are usually accompanied by a leukocytosis of 17,000 to 30,000 with neutrophilia and a release of band cells in a typical acute response to bacterial infection.[23] A WBC count of 27,000 to 44,000 was reported in a porpoise with a jaw abscess,[17] and another porpoise with an abdominal abscess had a WBC of 60,000.[21]

The reported normal blood parameters tend to vary widely among individual animals within a species, and between reports of averages within a single species. WBC counts up to 16,100 have been reported in normal young grey seals.[24] A one year survey of leukocyte counts in young captive harp seals demonstrated periodic fluctuation from a mean low of 4,000 to a mean high of 12,500 in healthy animals.[25] Fluctuations in blood parameters may be due to individual sampling techniques and induced stress, endogenous stresses of captive or feral conditions, or to normal periodicity associated with sexual cycling, molting, or other physiological factors that may be more vital in marine mammals than in the mammals that have been more extensively studied. The stresses of shipping can increase the leukocyte count of porpoises 30 to 40%, with a typical adrenalcorticosteroid response of neutrophilia and eosinopenia, but without lymphopenia.[21,26]

III. SERUM CHEMISTRIES

The blood urea nitrogen levels are somewhat higher in marine mammals than in terrestrial mammals, and can range from 18 to 95 as normal values reported in different species (Table 2). The cause of these normal elevated levels has not been discussed but it may be partially or entirely due to the natural high protein diet and amino acid catabolism. Creatinine levels will rise after a fish meal, due to protein catabolism.[17] The BUN tends to be slightly to moderately elevated as a prerenal uremia in some stranded pinnipeds that are emaciated and dehydrated.

Subadult elephant seals have high normal serum alkaline phosphatase levels, of bone origin, in these rapidly growing animals.[27] Alkaline phosphatase activity in small cetaceans is highest in the kidney, but renal disease may not elevate the serum levels since the enzyme may pass into the urine. Elevated alkaline phosphatase levels in the serum of marine mammals have not been linked to diseases of the biliary tract, as in other animal species.[28]

The most specific enzymes to measure liver disease in marine mammals are serum glutamic pyruvate transaminase (SGPT) and sorbitol dehydrogenase (SDH).[28,29] Serum glutamic oxalic transaminase (SGOT) is also an indicator of liver damage but this enzyme is widely distributed in the body, especially muscle, hence is less specific, as in many other animals.

The following tables list some of the reported normal values of serum enzymes and chemistries in marine mammals. The reported enzyme levels are confusing because limited numbers of animals have been tested for most marine mammal species, and the values are reported in various units. One would be advised to establish normal values with a laboratory for the species under consideration. However, as a guideline, the normal canine serum enzyme levels usually appear to be comparable to normal marine mammals.

Clinically, hypoglycemia, characterized by periodic short convulsions, in recently stranded young pinnipeds has been recognized.[23,27] These animals are very emaciated at the time of stranding and they therefore have a severe caloric deficit. Frequent feeding of these animals may be needed to keep the blood glucose levels elevated to the proper range, until body fat stores are replenished.

Table 1
HEMATOLOGY OF SELECTED MARINE MAMMALS[a]

Species	PCV	Hb g/dℓ	RBC × 10⁶/μℓ	MCV	MCHC g/dℓ	WBC	Segs	Bands	L	M	E
Bottle-nose dolphin	41—49	14—16	3.5—4	113—124	32—35	6—12,000	55—65	1—5	15—25	1—5	6—27
Killer whale	43—49	15—17	4	111—119	33—36	6—9,000	50—75	2—10	15—30	1—5	2—8
Harbor seal	50—60	18—32	5—6	117[b]	36.5[b]	7—9,000	38—72	1—2	17—51	2—7	1—11
Elephant seal	59—69	23—26	4—6			7700—12,000[c]	44—66	0	24—52	2—4	3—5
California sea lion	40—50	15	4—5	103	35[b]	7700—10,700	51—65	0—6	18—38	2—6	0—8

[a] Adapted from Medway and Geraci,[21] 1978, except as noted.
[b] Calculated from average data in Reference 31.
[c] Range 6 to 9,000 in healthy 1 to 2-year-olds at Marineland, Palos Verdes, Ca.[23]

Table 2
NORMAL BLOOD CHEMISTRY OF SELECTED MARINE MAMMALS[a]

	mEg/L			mg/dℓ			mU/mℓ				g/dℓ	
Species	Na	K	Cl	Ca	BUN	Glucose	SGOT[b]	SGPT[b]	A.P.[c]	S.D.[d]	Total Protein	Albumin
Bottle-nose dolphin	150—161	4—5	104—118	7—8	41—66	95—155	33—190	7—19	33—403	29—435	7—8	3—5
Killer whale	149—158	4	108—119	8—9	33[e]	203[e]					7—8	4
Harbor seal	147—156	4—5	100—110	5—6	34—59	121—152					7—8	3—4
Harp seal	147—161	4—6	90—117	6—12	20—69	104—170	11—61	5—12		58—220		
California sea lion	143—148	3—4	105—115		18—40	107—167	15—70[f]	20—60	2—18		7—9	3—4

[a] Adapted from Medway and Geraci,[21] 1978, except as noted, (plasma levels).
[b] Serum glutamic oxaloacetic transaminase and serum glutamic pyruvic transaminase, international units.
[c] Alkaline phosphatase, international units.
[d] Sorbital dehydrogenase, sigma units/mℓ.
[e] Reference 30.
[f] Reference 31.

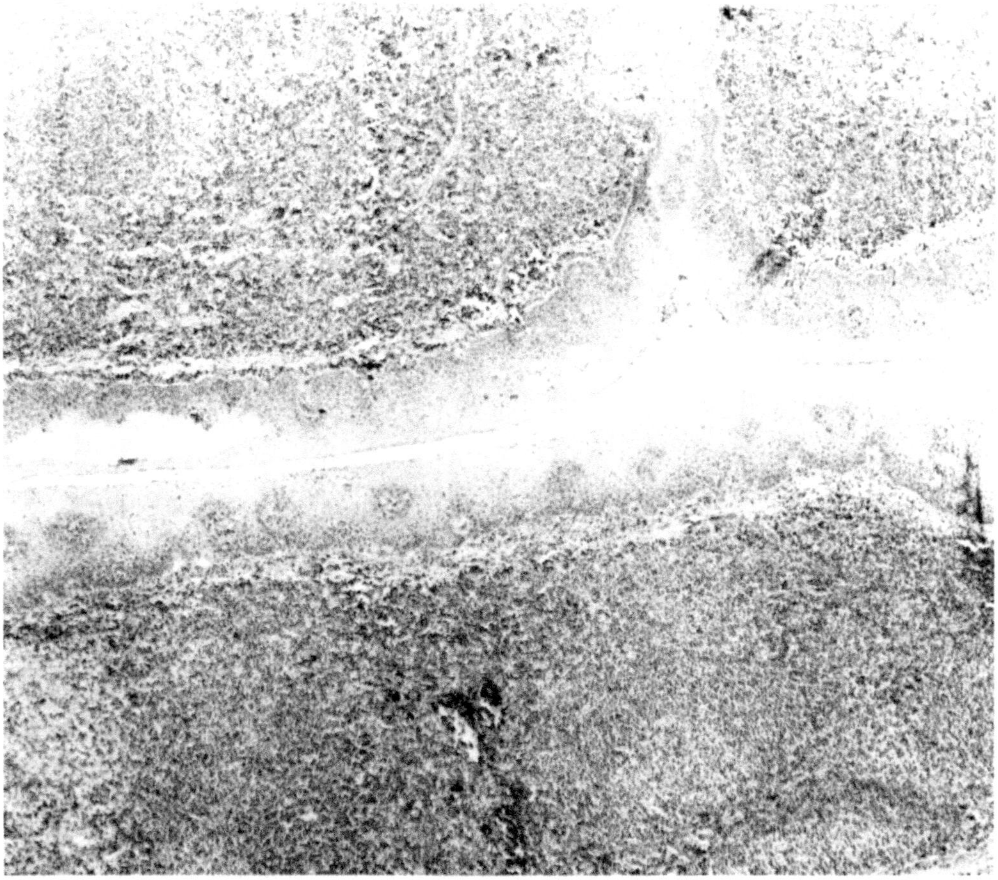

FIGURE 1. Tonsillar crypt with inactive follicles in a California sea lion, estimated to be 18 months old.

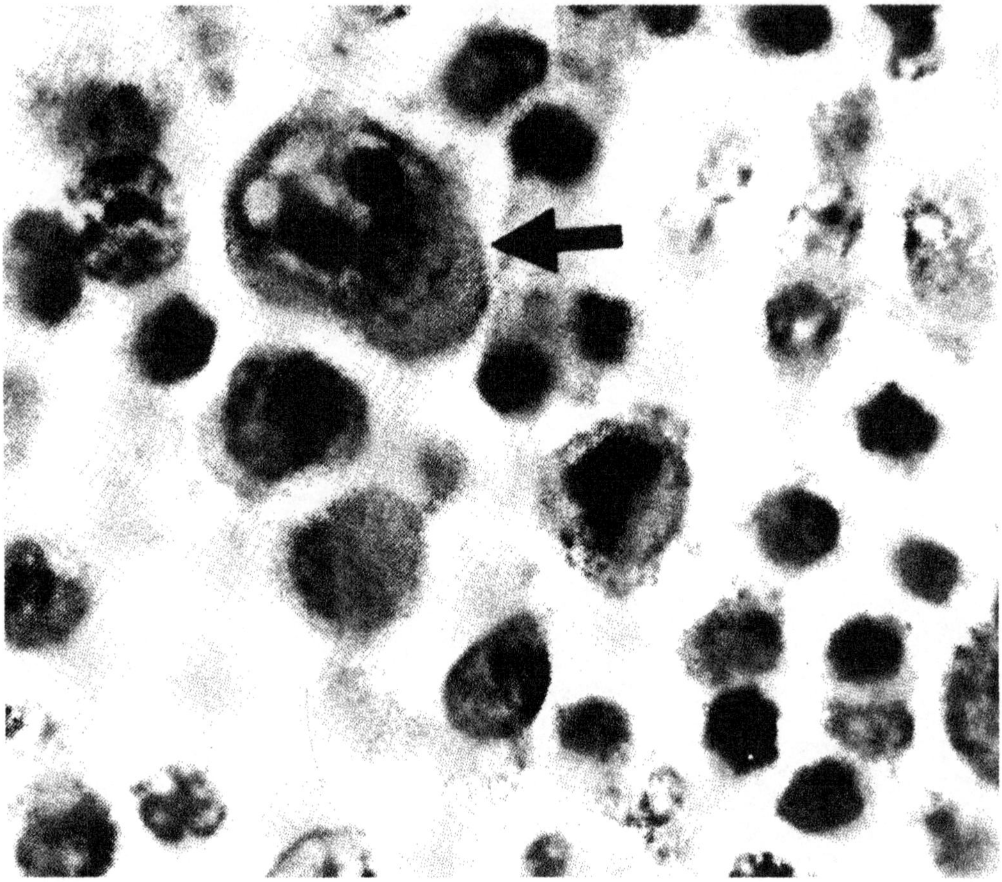

FIGURE 2. Reactive mesenteric lymph node with histiocytes and a binucleate Reed-Sternberg-like cell (arrow) in a northern fur seal pup infected with hookworms (Courtesy of J.A.V.M.A.).

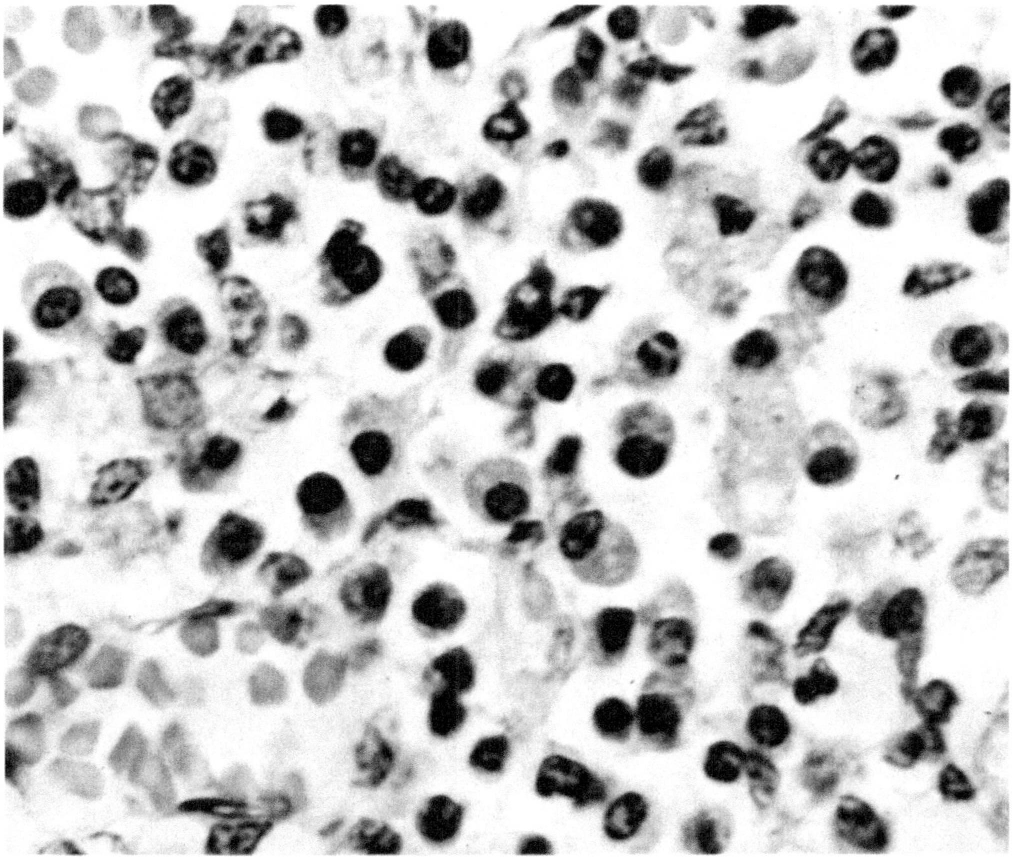

FIGURE 3. Plasmacytosis in a bronchial lymph node of a California sea lion with pneumonia.

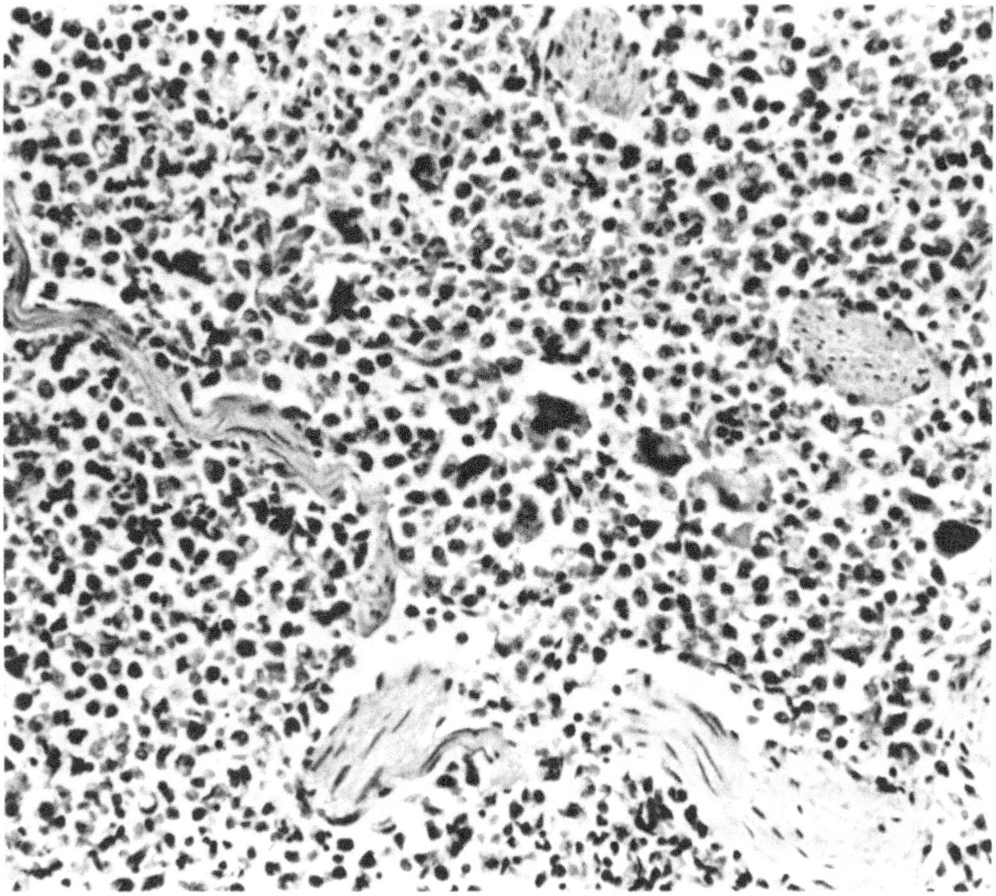

FIGURE 4. Myeloid metaplasia in the spleen of a California sea lion. Scattered megakaryocytes are seen.

FIGURE 5. Gross photograph showing the dark, round spleen and a single small accessory spleen in a common dolphin.

REFERENCES

1. **Simpson, J. G. and Gardner, M. B.,** Comparative microscopic anatomy of selected marine mammals, in *Mammals of the Sea,* Ridgway, S. H., Ed., Charles C Thomas, Springfield, Ill., 1972, 298.
2. **Cavagnolo, R. Z.,** The immunology of marine mammals, *Devel. Comp. Immunol.,* 3, 245, 1979.
3. **Uys, C. J. and Best, P. B.,** Pathology of lesions observed in whales flensed at Saldanha Bay, South Africa, *J. Comp. Pathol.,* 76, 407, 1966.
4. **Brown, R. J., Smith, A. W., Keyes, M. C., Trevethan, W. P., and Kupper, J. L.,** Lesions associated with fatal hookworm infections in the northern fur seal, *J.A.V.M.A.,* 165, 804, 1974.
5. **Mumford, D. M.,** Lymphocyte transformation studies of sea mammal blood, *Experientia,* 31, 498, 1975.
6. **Nash, D. R. and Mack, J.-P.,** Immunoglobulin classes in aquatic mammals characterized by serologic cross-reactivity, molecular size and binding of human free secretory component, *J. Immunol.,* 107, 1424, 1977.
7. **Mach, J.-P.,** In vitro combination of human and bovine free secretory component with IgA of various species, *Nature (London),* 228, 1278, 1970.
8. **Cavagnolo, R. Z. and Vedros, N. A.,** Identification and characterization of three immunoglobulin classes in the northern fur seal, *Callorhinus ursinus, Devel. Comp. Immunol.,* 2, 689, 1978.
9. **Olsen, O. W. and Lyon, E. T.,** Life cycle of *Uncinaria lucasi* stiles, 1901 (Nematoda: Ancylostomatidae) of fur seals, *Callorhinus ursinus* Linn., in the Pribilof Islands, Alaska, *J. Parasitol.,* 51, 689, 1965.
10. **Cavagnolo, R. Z. and Vedros, N. A.,** Serum and colostrum immunoglobulin levels in the northern fur seal, *Callorhinus ursinus, Devel. Comp. Immunol.,* 3, 139, 1979.
11. **Landy, R. B.,** A review of neoplasia in marine mammals (Pinnipedia and Cetacea) in *Comparative Pathology of Zoo Animals,* Montali, R. J., and Migaki, G., Eds., Smithsoian Institution Press, Washington, D.C., 1980, 579.
12. **Griner, L. A.,** Malignant leukemic lymphoma in two harbor seals *(Phoca vitulina geronimensis), A.J.V.R.,* 32, 827, 1971.
13. **Stedham, M. A., Casey, H. W., and Keyes, M. C.,** Lymphosarcoma in an infant fur seal (*Callorhinus ursinus*), *J. Wildlife Dis.,* 13, 176, 1977.
14. **Smith, H. A., Jones, T. C., and Hunt, R. D.,** *Veterinary Pathology,* 4th Ed., Lea & Febiger, Philadelphia, 1972, 187.
15. **Clamen, H. N.,** Corticosteroids and lymphoid cells, *N. Engl. J. Med.,* 287, 388, 1972.
16. **Arvy, L. and Pilleri, G.,** Some characteristics of the cetacean spleen, *Invest. Cetacea,* 2, 165, 1970.
17. **Ridgway, S. H.,** Homeostasis in the aquatic environment, in *Mammals of the Sea,* Ridgway, S. H., Ed., Charles C Thomas, Springfield, Ill., 1972, 590.
18. **Medway, W.,** Iron metabolism and some of its aspects in captive cetaceans, in *Proc. 2nd Symposium on Diseases and Husbandry of Aquatic Mammals,* Caldwell, D. K. and Caldwell, M. C., Eds., Marineland of Florida, St. Augustine, 1968, 123.
19. **Andersen, S.,** The physiological range of formed elements in the blood of the harbor porpoise, *Phocaena phocaena, Nord. Vet. Med.,* 8, 51, 1966.
20. **Anderson, S.,** personal communication, 1967, cited by Reference #17.
21. **Medway, W. and Geraci, J. R.,** Clinical pathology of marine mammals, in *Zoo and Wild Animal Medicine,* Fowler, M. E., Ed., W. B. Saunders, Philadelphia, 1978, 604.
22. **Fleishman, R. W. and Squires, R. A.,** Verminous pneumonia in the California sea lion *(Zalophus californianus), Vet. Pathol.,* 7, 89, 1970.
23. **Andrews, B. and Sweeney, J. C.,** personal communications, 1981.
24. **Greenwood, A. G., Ridgway, S. H., and Harrison, R. J.,** Blood values in young gray seals, *J.A.V.M.A.,* 159, 571, 1971.
25. **Ronald, K., Foster, M. E., and Johnson, E.,** The harp seal (*Pagophilus groenlandicus*). II. Physical blood properties, *Can. J. Zool.,* 47, 461, 1969.
26. **Medway, W. and Geraci, J. R.,** Hematology of the bottle-nose dolphin *(Tursiops truncatus), Am. J. Physiol.,* 207, 1367, 1964.
27. **MacDonald, K.,** personal communication, 1982.
28. **Geraci, J. R. and St. Aubin, D. J.,** Tissue sources and diagnostic value of circulating enzymes in cetaceans, *J. Fisheries Res. Bd. (Canada),* 36, 158, 1979.
29. **St. Aubin, D. J. and Geraci, J. R.,** Tissue distribution and plasma levels of liver enzymes in phocid seals, *Can. J. Zool.,* 55, 1936, 1977.
30. **Ridgway, S. H., Simpson, J. G., Gilmartin, W. G., and Patton, G. S.,** Hematological findings in some small cetaceans, *J.A.V.M.A.,* 157, 560, 1970.
31. **Sweeney, J. C.,** Procedures for clinical management of pinnipeds, *J.A.V.M.A.,* 165, 811, 1974.

Filtration, disinfection, and circulation of aquarium water often proves inadequate to maintain an environment such that these animals do not develop mycotic or bacterial infections. Attempts have been made to further purify the water in such a closed system by employing ultraviolet (UV) irradiation of the water on the effluent side of the filter system,[32] but because of the large volume of water involved and the relative inefficiency of UV radiation, the system was found to have little value in reducing bacterial counts in the various tanks. One might wonder if a gamma-irradiation source would be more effective, but then the additional requirements for shielding and safety may be excessive.

We have wondered about the possible effects of the environmental pollution encountered by these animals in the offshore waters of locations adjacent to large urban areas. It is conceivable that there is an insidious, unmeasured effect on these animals, which results in a greater susceptibility to bacterial, mycotic, and viral infections. We have described the fact that there is a significant offshore pollution in the Los Angeles coastal waters, particularly with the organochlorine compounds, and in this book, the tissue residues of some of these compounds is documented.

Numerous studies have documented the effects of DDT type organochlorine insecticides on the immune system. It has been shown that there is a decreased antibody response to exogenous antigen, a decrease in total circulating immunoglobulin levels, and decrease in splenic lymphoid mass.[33]

The effects of polychlorinated biphenyl (PCB) compounds on the immune system are much more striking than DDT causes, and have been noted in various species up to and including man.[34] Not only is there atrophy of lymphoid tissue and decreased immunoglobulin levels, but in both experimental animals and accidentally exposed humans, there is increased susceptibility to viral, bacterial, and mycotic infections. Chronic respiratory infections and pneumonias were a common observation in man and a common cause of death in experimentally exposed primates and rats.

The effects of these organochlorine type compounds on the immune system is magnified if the system is exposed in the formative perinatal period of life. This is the situation which exists with these marine mammals, particularly the pinnipeds, who bear and rear their offspring in these same contaminated offshore waters. There has been evidence of a link between the contamination and premature or underweight sea lion pups off southern California.[35] A careful and responsible study should be conducted to help determine if these animals have an impaired immune system and may have a greater susceptibility to bacterial, viral, or mycotic diseases. Lymphocytic function studies, as well as response to exogenous antigen and macrophage functional status could help to determine if such an effect is present in these feral animals, using an appropriate noncontaminated animal group for control data.

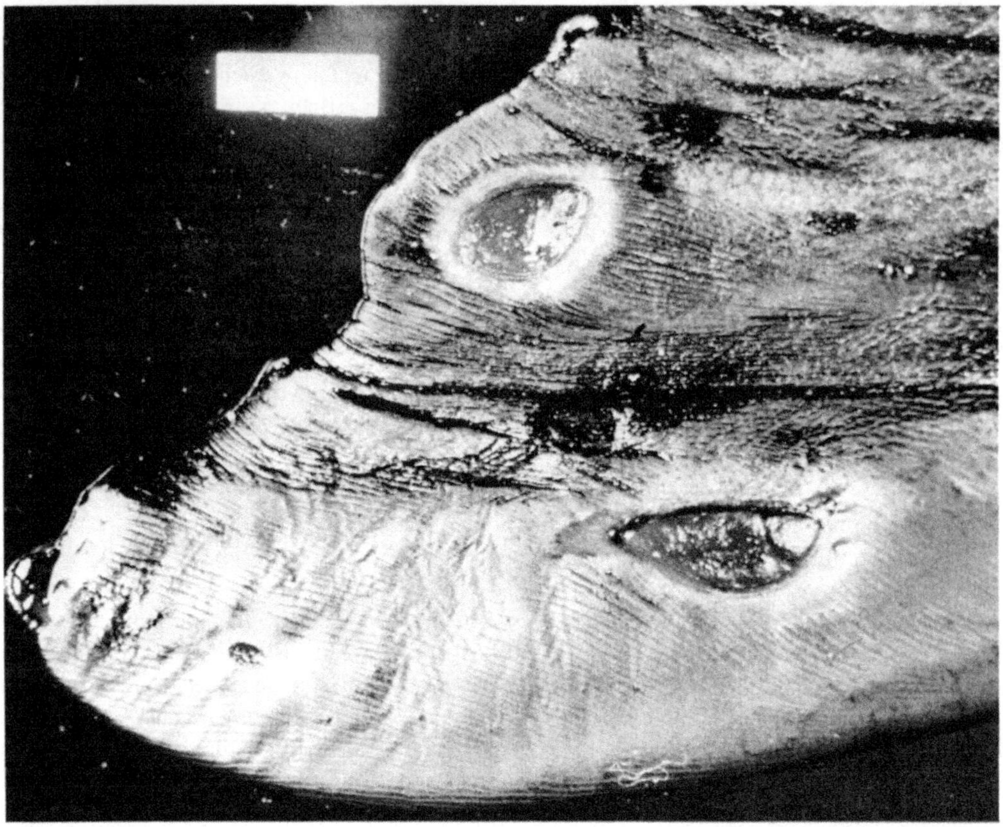

FIGURE 1. This is a flipper from a California sea lion with two chronic bacteria-associated discrete crater-like ulcers. Trauma could have been the initiating factor.

the effects of organochlorines on the immune system in laboratory animals.[9] Few tests have been done to specifically measure the effects, if any, on the cell mediated immune system. Tests to evaluate the humoral immune system have been extremely variable in results obtained, but usually have shown that DDT has no detrimental effect on humoral immunity, as determined by serum gamma-globulin levels and antibody response to various injected antigens. Koller concluded that consistent effects of DDT in humoral immunity appear to be minimal, based on conflicting results of various investigators. Toxicity studies have not been commonly done on DDE, the primary metabolite of DDT.

The effects of PCB's on the immune system were also reviewed,[9] and the various reports describe decreases in primary and/or secondary immune response to exogenous antigens, decrease in total serum gamma-globulin, increased susceptibility to viral infections, and decreased germinal center activity in lymph nodes. It is interesting that the presence of active lymphoid germinal centers in nodes of northern fur seals of the Bering Sea,[10] and the relative absence of germinal center activity of lymphoid tissue of California sea lions, are correlated with low DDT liver concentrations in the northern fur seal (average 0.45 ppm),[11] while there are high DDT levels in the livers of the sea lions, according to our data and that of others, who reported 17 ppm or more.[9] Additionally, practically no PCB's were detected in the fur seals, while our data show that high levels of PCB's in the livers of California sea lions are a common finding.

Organochlorine compounds are highly fat soluble. Marine mammals ingest these compounds in the tissues of contaminated food fish. Most of the ingested DDT compounds are in the form of DDE, which accounts for over 80% of the total DDT group. DDE shows less acute toxicity, and has no insecticidal activity. Some of the DDT that is ingested is metabolized and excreted before it can be distributed throughout the body.[4]

One study of California sea lions along the Oregon coast compared the distribution of DDT and its metabolites in tissues in 1971 and again in 1973.[12] No DDT was detected in 1973, only DDD and DDE, suggesting that no new DDT had been introduced into the environment since the previous test period. DDT had been banned for general use in 1970. Other studies have found residues of unmetabolized DDT after 1973 however, and our data have shown continued DDT residues in some animals in 1980.

The blood level of total DDT, or of PCB, does not exceed 1 to 3% of the stored fat tissue levels in seals,[13] and in humans it has been determined that the fat to blood ratio of DDT is about 300:1.[14,15] Plasma has about twice the level of DDT as whole blood, in humans.[15]

In marine mammals, these organochlorine compounds are stored in blubber, because this is the principal fatty tissue of these animals. The next most concentrated tissue for these lipid phase compounds is the liver, where tissue levels are generally 5 to 20% of the fat levels. Pinniped milk also has significant levels of organochlorine compounds, since it consists of 50% fat in fur seals[13] (Tables 1 and 2).

For some unknown reason, DDT and PCB do not tend to accumulate in brain tissue. The brain tissue of harp seal pups has 8.3% extractable lipid, compared to 3.4 to 4.0% for the liver, but the organochlorine concentrations are lower in the brain than in the liver.[1] This suggests that there may be a partial blood-brain barrier which tends to exclude these compounds. Also, it may be that the liver levels are higher than other organs because of the transport of organochlorines attached to chylomicrons entering the liver from the intestine, for metabolism and redistribution.

The eventual fate of PCB's and DDT's stored in the blubber is poorly understood, but is important in understanding the mechanism of possible toxicity and modes of excretion. In fasting birds, that are catabolizing their fat stores, the DDT is redistributed in the body and builds up to toxic levels even in the absence of further exposure.[8]

The liver mixed function oxidase enzymes are stimulated to increase the rate of excretion of organochlorines, but the mechanism appears to be ineffective compared to the rate of

Table 1
LITERATURE REVIEW: ORGANOCHLORINE LEVELS IN PINNIPEDS—MEAN LEVEL IN FAT (PPM)

Species	Location	Number	Year	PCB	Total DDT	Dieldrin	Ref.
Gray/harbor seals	Eastern Scotland	18	1967		14.5	0.79	34
	N. & W. Scotland	6	1967		7.61	0.20	34
	Canada	2	1967		1.68	0.03	34
Gray seals	Canada	10	1977	14.5	14.0		16
Gray seal pups	Canada	10	1977	2.2	7.1		16
California sea lions							
Healthy	Oregon	9	1976	29.8	283 (DDE)		32
Sick	Oregon	9	1976	25.2	384 (DDE)		32
California sea lion cows							
With premature pups	Southern California	6	1973	112.4	824.4	0.0	19
With full-term pups	Southern California	4	1973	17.1	103.2	0.0	19
California sea lions	Northern California	25	1971		911.0		9
Northern fur seals							
Dams	Alaska	2	1972	5.8	5.2	0.11	13
Pups	Alaska	3	1972	33.0	63.0	0.02	13
Northern fur seals	Alaska	30	1970	0.0	0.45 (liver)		12
Bearded seals	Arctic	5	1967	1.8	0.47 (DDE)		35
Ringed seals	Baltic	85	1976	94.7	115.0		17
Ringed seal fetuses	Baltic	24	1976	49.0	62.0		17
Ringed seals	Arctic	5	1967	0.9	0.026 (DDE)		35
Hooded seals	Arctic	5	1967	2.7	0.29 (DDE)		35
Harp seals	Canada	18	1973	2—22	3.1—22.6	0.1—0.3	36
Harp seals	Canada	11	1973	3.231	4.02	0.81	1
Harp seal pups	Canada	20	1973	2.263	1.981	0.93	1
Monk seal	Hawaii	1	1981	10.6	11.73 (DDE)		37

Table 2
ORGANOCHLORINE LEVELS IN CETACEANS (REVIEW) PPM IN FAT, RANGE OR MEAN

Species	Number	Year	PCB	Total DDT	Dieldrin	Ref.
Gray whale	23	1970		0—0.68	0—0.075	38
Sperm whale	6	1970		1.82—9.43	0—0.19	38
Sperm whale	3	1975	0.7—4.0	1.1—15.0	0	39
Fin whale	12	1973	0.12—0.185	0.667—2.557	0	40
Short-finned pilot whale	5	1974	0.69—1.60	1.25—2.28	0.01—0.04	41
Atlantic pilot whale	2	1975	42 & 114	30.3 & 268	1.1 & 3.0	39
Pacific pilot whale	1	1975	46	255.3	0.4	39
Pilot whale	1	1971		68.8		42
Dense-beaked whale	2	1975	14 & 29	38.2 & 65.1	0.2 & 0.3	39
Hump-back whale	4	1975	1.3—6.0	1.4—23.1	0—0.1	39
Beluga whale	14	1973		3.23		43
Atlantic bottle-nose dolphin	2	1972	29 & 41	13.9 & 26.9	0.049 & 0.065	44
Long-snouted dolphin	2	1974	2.0 & 5.0	1.44 & 7.38	0.007 & 0.05	41
Common dolphin	3	1972	0.75	4.27—177	0.021—0.25	44
Common dolphin	1	1975	0.69	70.9	2.6	39
Harbor porpoise	3	1967		42.8	9.9	34
Harbor porpoise	7	1972	35—148	11.1—102	0.013—0.54	44
Harbor porpoise	1	1975	74	57.5	1.5	39
Harbor porpoise	2	1967	6.7	0.32		35
Striped dolphin	2	1975	39 & 69	70.7 & 231	1.4 & 2.4	39
Pacific white-sided dolphin (captive)	1	1975	147	1023	4.1	39

ingestion, based on reports of PCB and DDT levels in various marine mammal populations. Some PCB's have a tissue half-life in rats which is greater than the life span of the animal,[15] while other PCB's have a half-life of only one day or so. Levels of stored organochlorines in the fat may reach a maximum level and stabilize, even though the pollutants continue to be ingested. This is a normal pharmacologic effect with continued ingestion of compounds that are not acutely toxic. For example, rats fed 220 ppm DDE in their diet reach a maximum fat level of 800 ppm in males and 2000 ppm in females.[14]

Most of the pinnipeds we examined in southern California were extremely emaciated at the time of death, due to chronic lungworm and stomach worm infection, as well as for unknown reasons. The majority of these animals have no blubber layer to sample. The high liver concentrations of total DDT and PCB we have measured (Tables 3—8) could be due to the redistribution of the fat soluble organochlorines to the blood and liver and other tissues during periods of fat catabolism. The levels of these compounds could reach toxic levels as the blubber is metabolized, as occurs in starving birds, but this possibility has apparently not been studied. Clinically, we often see starving young sea lions with normal blood glucose levels that go into periodic convulsions for no detectable reason, that could be associated with organochlorine toxicity. Possibly the many emaciated stranded pinnipeds without any obvious reason for their poor nutritional condition are suffering from a primary toxicosis due to organochlorines or other pollutants. This would seem unlikely because the intake of various pollutants should average out between our local sea lions eating a large number of various fish, and the vast majority of animals appears to be healthy.

Other than by slow metabolism, PCB and DDT levels can only be lowered in adults by maternal transfer to the fetus and by lactation.[1,13,16] Organochlorine levels increase with age in male pinnipeds according to some studies, but not in females.[16] One interesting study measured DDT and PCB levels in grey seal milk and calculated that during the lactation period the dam would lose 30% of her total DDT and 15% of her total PCB stores, balancing the annual dietary intake.[16] The milk lipid has only about 60% and 30% of the dam's blubber levels of DDT and PCB respectively. Although still high, it was not known why these compounds were somewhat less concentrated in milk fat, as compared to body fat. The net effect of this lactogenic chlorinated hydrocarbon transfer to the pups is that the fat stores of DDT and PCB are much greater in 2-month-old pups than in their dams. The cows' fish diet has less concentrated residue levels than does the milk they are producing.[13] The limited study of two cows and three pups found a mean fat level of 5.8 ppm of PCB in the cows and 33 ppm in the 2-month-old pups, and respective levels of 5.2 ppm and 63 ppm of DDT. They cited other workers who did not find these differences in other pinniped species. Possibly, there was a sampling error in such a small study, or degradation rates may vary for different species.

Residues are transferred in utero,[1] but they are not concentrated in the fetus compared to the dam.[13,17] This is consistent with the observation that neonatal human babies also have low level DDT and PCB residues.[18] Organochlorines have been incriminated as possible causes of abortion, neonatal death, infertility, and population declines in some groups of pinnipeds. The low reproductive rate of seals in the highly polluted Baltic Sea has been suggested to be due to DDT and PCB pollution. One study surveyed 109 female and 24 male ringed seals of reproductive age in the Baltic Sea.[17] Of these, there were 47 nonpregnant females with segmental stenosis and occlusion of the lumen of the uterus. The lumens contained an exudate that was usually reddish and watery, but sometimes was turbid or mustard-like in texture and color. These changes were also noted in about one-half of the limited number of grey and harbor seals examined but they did not determine the nature of the exudate or uterine lesions. They did not know if the uterine lesions were directly associated with the levels of PCB and DDT, but the residues were significantly higher in those animals than in pregnant females (nonpregnant seals with uterine stenosis had mean fat DDT and

Table 3
ORGANOCHLORINE RESIDUES (PPM) IN SEALS (SUB-ADULT)

Animal #	Sex	Year	Necropsy findings	PCB	DDT	TDE	DDE	Total DDT
			Mirounga angustirostris (Northern Elephant Seal)					
56416	M	1974	Peritonitis (pup)	0	0	0.12	0.16	0.28
80663		1979	Emaciation	2.5	1.2	0	12.0	13.2
85625	M	1979	Verminous pneumonia	3.2	0	0	9.6	9.6
85629	M	1979	Emaciation	9.0	0	0	15.0	15.0
86341	M	1980	None	2.6	0.77	0	44.0	44.77
86943	F	1980	Enteritis	0.56	0	0	3.4	3.4
90775	F	1981	Pneumonia & nephrosis	2.5	0.16	0	51.0	51.16
			Phoca vitulina (Harbor Seal)					
66178	M	1976	Septicemia	14.0	0	0	8.7	8.7
83005	M	1979	Panophthalmitis	0.18	0	0	11.0	11.0
83985	M	1979	Enteritis	11.0	0	9	92.0	101.0
88562	F	1980	Verminous pneumonia	7.1	1.3	1.0	26.0	28.3
			Callorhinus ursinus (Northern Fur Seal)					
85483	F	1979	Gastritis & abscess	9.7	0	2.7	12.5	15.2
86001	M	1980	None	5.9	0.73	1.5	12.9	15.13
86901	F	1980	Toxoplasmosis	1.9	0	0	11.1	11.1
87324	F	1980	Enteritis	3.7	0.30	0.61	3.9	4.81
			Liver tissue — wet wt. $\bar{X}$	4.9	0.297	0.395	20.88	22.176

Table 4
ORGANOCHLORINE RESIDUES (PPM) IN CALIFORNIA SEA LIONS (PREMATURE/STILLBORN)

Animal #	Sex	Year	PCB	DDT	TDE	DDE	Total DDT
66090		1976	0.6	0	0	1.3	1.3
66203	M	1976	0.92	0	0	0.54	0.54
82032	F	1979	0.4	0.13	0.079	0.91	1.12
87866	M	1980	0.5	0	0	0.15	0.15
Liver tissue — wet wt. $\bar{X}$			0.6	0.03	0.02	0.72	0.78

Note: The dams were in the captive breeding population at Marineland

Table 5
ORGANOCHLORINE RESIDUES (PPM) IN CALIFORNIA SEA LIONS (YOUNG MALES — 1 TO 24 MONTHS)

Animal #	Year	Necropsy findings	PCB	DDT	TDE	DDE	Total DDT
51940	1973	Verminous pneumonia	15.0	3.8	11.0	111.0	125.8
66029	1976	Bacterial pneumonia		0.3	6.6	21.0	27.9
71803	1977	Bacterial pneumonia	67.0	26.0	35.0	439.0	500.0
72586	1977	Pneumonia & peritonitis	87.0	0	0.98	0.27	1.25
80601	1979	Enteritis	25.0	21.0	11.0	123.0	155.0

Table 5 (continued)
ORGANOCHLORINE RESIDUES (PPM) IN CALIFORNIA SEA LIONS (YOUNG MALES — 1 TO 24 MONTHS)

Animal #	Year	Necropsy findings	PCB	DDT	TDE	DDE	Total DDT
85788	1980	Pneumonia & peritonitis	28.0	22.0	8.5	118.0	148.5
86853	1980	Emaciation	19.0	3.1	5.6	36.0	44.7
87100	1980	Emaciation	11.0	0.9	1.6	9.0	11.5
87101	1980	Vascular collapse	0.67	0	0	0.48	0.48
87325	1980	Emaciation	1.1	0	0.075	3.0	3.075
87660	1980	Enteritis	3.5	0.2	0.61	22.0	22.81
87662	1980	Emaciation		0.15	1.8	8.1	10.05
87719	1980	Verminous pneumonia	3.3	0.19	0.086	22.0	22.276
87779	1980	Peritonitis	0.47	0	0	5.1	5.1
87871	1980	Verminous pneumonia	2.0	0.11	0.17	1.0	1.28
88101	1980	Verminous gastritis	16.0	6.5	6.9	117.0	130.4
88106	1980	Collapsed lung	5.0	0.42	0.55	5.6	6.57
92915	1981	Verminous pneumonia	3.4	0	0.49	1.6	2.09
		Liver level — wet wt. $\overline{X}$	16.0	4.7	5.1	58.0	67.8

Table 6
ORGANOCHLORINE RESIDUES (PPM) IN CALIFORNIA SEA LIONS (YOUNG FEMALES — 1 TO 24 MONTHS)

Animal #	Year	Necropsy Findings	PCB	DDT	TDE	DDE	Total DDT
32956	1970	Leptospirosis	90.0	23.0	45.0	366.0	434.0
55836	1974	Skull fracture	0.95	0.05	0.16	3.8	4.01
62848	1975	Verminous pneumonia		3.0	7.2	47.0	57.2
65193	1976	Emaciation	1.1	0	0	7.6	7.6
72191	1977	Salmonellosis	25.0	2.3	4.8	32.0	39.1
76527	1978	Verminous pneumonia		0	0.23	1.8	2.03
81058	1979	Cellulitis	15.0	4.4	4.6	54.0	63.0
81851	1979	Emaciation	3.8	0.46	0.37	11.0	11.83
83988	1979	Pneumonia & enteritis	11.0	0	16.0	160.0	176.0
85530	1979	Abscesses	8.9	0	0	9.8	9.8
85627	1979	Pneumonia & gastritis	6.5	0	0.94	10.0	10.94
85630	1979	None	8.4	0	0	24.0	24.0
86400	1980	Meningoencephalitis	8.3	0	0.6	15.0	15.6
86806	1980	Pneumonia	8.9	4.4	5.7	48.0	58.1
86852	1980	Pneumonia	1.6	0.37	0.23	4.0	4.6
86902	1980	None	1.7	0.14	0.17	3.0	3.31
86942	1980	Parasitic gastroenteritis	6.2	3.0	1.8	35.0	39.8
87605	1980	Septicemia	2.0	0	0.18	3.9	4.08
87758	1980	Pulmonary edema	1.8	0	0	1.3	1.3
87762	1980	Emaciation	30.0	2.6	11.0	64.0	77.6
88333	1980	Emaciation	5.2	0	0.14	22.0	22.14
		Liver level — wet wt. $\overline{X}$	11.3	2.1	5.1	43.6	50.8

Table 7
ORGANOCHLORINE RESIDUES (PPM) IN CALIFORNIA SEA LIONS (ADULT MALES)

Animal #	Year	Necropsy findings	PCB	DDT	TDE	DDE	Total DDT
33203	1970	None	2.2	0	7.2	36.0	43.2
70373	1977	Verminous pneumonia	55.0	0	4.4	164.0	168.4
82524	1979	Peritonitis		0.46	0.32	6.4	7.18
83383	1979	Pneumonia	0.8	0.34	0.08	1.59	2.01
84289	1979	Pneumonia & hepatosis	15.0	0	8.1	200.0	208.1
85274	1979	Pneumonia	59.0	52.0	12.0	181.0	245.0
85788	1980	Pneumonia & peritonitis	28.0	22.0	8.5	118.0	148.5
86992	1980	Pneumonia	3.3	0.14	0.29	4.7	5.13
87237	1980	Bile duct adenocarcinoma	6.9	0.39	0.72	34.0	35.11
87847	1980	Septicemia	3.3				
		Liver level — wet wt. $\overline{X}$	17.4	7.5	4.2	74.6	86.3

Table 8
ORGANOCHLORINE RESIDUES (PPM) IN CALIFORNIA SEA LIONS (ADULT FEMALES)

Animal #	Year	Necropsy findings	PCB	DDT	TDE	DDE	Total DDT
51939	1973	Parasitic liver fibrosis	9.4	0	2.0	67.0	69.0
57994	1975	Carcinoma, urinary bladder		0	0.56	0	0.56
59319	1975	Carcinoma, ovary		0	0	0	0
61732	1975	Leiomyoma, uterus pneumonia & gastric abscess		0	0.17	0	0.17
74463	1977	Verminous pneumonia		0	0.65	1.0	1.65
84175	1979	Adrenal atrophy	0.7	0	0	0.51	0.51
84477	1979	Pneumonia		0	0.9	35.0	35.9
85456	1979	Renal hypoplasia	25.0	0	14.0	139.5	153.5
85529	1979	Myocarditis	15.0	0	0	11.0	11.0
85626	1979	Interstitial nephritis	19.0	4.5	3.7	21.0	29.2
85628	1979	Pneumonia	23.0	0	8.8	103.0	111.8
85789	1980	Adenocarcinoma, uterus	1.7	1.2	0.6	10.0	11.8
87102	1980	Pneumonia	25.0	0.84	0	78.0	78.84
87603	1980	Adenocarcinoma, uterus	4.8	0	0	10.0	10.0
92885	1981	Adenocarcinoma, uterus	1.47	0.21	0.49	0.25	0.95
93437	1981	Hepatopathy	3.46	0	0	0	0
		Liver level — wet wt. $\overline{X}$	8.0	0.4	2.0	29.8	32.2

PCB levels of 130 ppm and 110 ppm respectively vs. 88 ppm and 73 ppm in pregnant animals). Normally breeding California sea lions and nonpregnant Baltic seals reportedly have similar DDT levels, but the levels of PCB are higher in the Baltic seals.[17,19]

Infertility in mink can be experimentally produced by feeding PCB's, causing increased fetal resorption.[17] Levels of PCB in the mink are similar to those found in Baltic seals, but increased DDT levels in mink are not related to fetal resorption. Our data on the following chart indicate that southern California sea lions in fact have higher liver concentrations of PCB's than previously reported.[19] None of our animals were pregnant, but we believe that their residue levels are indicative of the general exposure of animals in the rookeries, hence the causal relationship between PCB's and infertility is not yet established. Also, it is possible that the mature female ringed seals that were infertile had higher organochlorine tissue levels because they did not have the lactational route of residue elimination. It is unknown whether the uterine lesions were the cause of the infertility, a reaction to fetal resorption, or were due to secondary infection.

Studies on the feral sea lion rookery population off the southern California coast have also correlated premature whelping with increased levels of organochlorines.[19] Six females with premature pups and four females with full-term pups were collected within 24 hr after parturition. The average total DDT level in the blubber of the dams with premature pups was 824.4 ppm compared to 103.2 ppm in dams with full-term pups. Liver concentrations were also markedly different, 25.24 ppm in dams of premature pups, vs. 6.67 ppm in dams with full-term pups, and the premature pups had 2.38 ppm of DDT in the brain compared to 1.2 ppm in the mature neonates. PCB level differences were similar to DDT; the dams of the premature pups had an average of 112.4 ppm in blubber and 5.74 ppm in liver, compared to dams of full-term pups with blubber and liver levels of 17.1 and 1.32 ppm, respectively. Again, the PCB levels in the brains of the pups were low, but higher levels were in the brains of premature pups, 0.45 ppm vs. 0.19 ppm, reflecting the higher tissue levels of their dams.

The mechanism by which organochlorines may be responsible for prematurity and infertility is poorly understood. The ring structure of these molecules may be similar enough to those of endogenous sex steroid hormones to compete for binding sites on gonadal or uterine target cells. Some organochlorines and their metabolites may have estrogenic activity in themselves.[20,21] Also the organochlorines increase the activity of liver microsomal enzymes so that steroid hormones are metabolized faster. DDT increases the metabolism of cortisol to 6B-hydroxycortisol,[22] and pigeons treated with DDT or dieldrin have increased rates of hydroxylation of testosterone and progesterone in liver fractions.[23] The effect of these two residues is additive, suggesting that they stimulate separate enzymes. It is unknown whether the increased activity of liver enzymes results in lower circulation levels of steroid hormones in marine mammals, but if it does an effect on reproduction might be expected. In human patients DDT has been used to lower circulating levels of cortisol with Cushing's disease. Similarly, polybrominated biphenyls cause a decrease in the level of circulating progesterone in monkeys.[24] In birds, DDT toxicity is thought to cause thinning of the egg shell by one or more mechanisms. The increased rate of steroid metabolism could alter the circulating levels of regulatory sex hormones, or could nonspecifically hydroxylate vitamin D, causing an artificial vitamin D deficiency, preventing proper calcium deposition in bones for later use.[25] Also, DDE may inhibit the enzyme carbonic anhydrase in the shell gland, which is responsible for the deposition of calcium carbonate.[25] Possibly other enzyme systems could be inhibited in the tissues of contaminated mammals. Experimentally DDT has been shown to cause light microscopic changes in hepatocytes, characterized by hypertrophy of the cytoplasm, with staining differences, probably due to hypertrophy of the smooth endoplasmic reticulum and the microsomal enzyme systems.[26] DDT excess is also associated with decreased liver vitamin A content, and has been shown to cause a riboflavin deficiency by interacting with polyunsaturated fatty acid metabolism.[27] Proliferative gastroenteritis has not been seen in PCB contaminated marine mammals as in primates.[28] Other PCB related lesions in experimental monkeys include microscopic hyperplasia of the bile duct epithelium, and macroscopic enlargement of the common bile duct, testicular degeneration, and chloracne-like lesions as seen in man. We see sea lions with focal or patchy areas of alopecia and thickened skin, that microscopically have dysplastic hair follicles plugged with keratin, and with superficial hyperkeratosis that appears similar to chloracne. Also, we have seen an occasional sea lion with microscopic intrahepatic bile duct hyperplasia and an occasional sea lion has been seen with massive enlargement of the common bile duct to a saccular structure about 8 cm in diameter. These cases were not associated with liver fluke infection, and the cause was unknown. To our knowledge, spermatogenesis has not been correlated with organochlorine levels in pinnipeds. The male pinnipeds available along the coast are very rarely of breeding age.

Since the chlorinated hydrocarbons are so persistent and stable, they will remain as important contaminants in the foreseeable future. Understanding the action of these residues in marine mammals will aid in evaluating possible effects in exposed humans.

II. HEAVY METALS

Since the unfortunate occurrence of human mercury poisoning in the Minamata Bay, Japan, episode of 1953 to 1960,[29] scientists have become especially aware of the potential dangers of this and other heavy metals in the environment. Inorganic mercury that is discharged into water from industry is transformed into toxic methylmercury by microorganisms,[30] and is then concentrated in fish and subsequently the mammals eating the fish. Methylmercury is toxic because it is readily absorbed from the intestinal tract and can pass the blood-brain barrier, accumulating in the brain, as well as in the kidneys and liver.[30] Average liver levels of total mercury as high as 95 ppm have been reported in apparently healthy California sea lions.[31] Encephalopathy, nephrosis, or other toxic effects have not been associated with mercury poisoning in marine mammals. Liver levels of total mercury in sea lion studied was a mean 82.12 ppm in healthy animals, and 161.6 ppm in sick animals. In the Minamata poisoning, the liver concentration of mercury was 22 to 70 ppm in 12 patients that died acutely, within 100 days of first illness. The liver mercury in the poisoned humans was primarily methylmercury, while the total liver mercury in the pinnipeds studied was less than 15% methylated, usually less than 2 to 3%,[31,32] even though the mercury in the fish diet is primarily methylmercury. The brain total mercury in those sea lions was low, about 0.5 ppm. The ingested methylmercury may be very efficiently demethylated in the pinniped liver before it can be distributed throughout the body.[31,32] Methylmercury experimentally increases selenium uptake,[30] and these pinnipeds had a 1:1 mercury/selenium ratio in the liver. The presence of selenium and its availability in the marine environment may somehow aid in the demethylation and detoxification of mercury.[32] Selenium containing compounds are converted in the body to selenide which reacts with mercury and methylmercury (and cadmium) cations to alter the distribution and toxicity of these meavy metals, and it may increase the rate of mercury excretion.[33] Pinniped populations closer to centers of industrial pollution accumulate mercury at a faster rate than animals living in more isolated areas because the fish are more contaminated.[32]

Cadmium has been found concentrated in the kidney, in several studies of pinnipeds. Sea lions from Oregon had an average kidney cadmium level of about 10 ppm, and in a review by Buhler et al.[31] there were findings of 22 ppm in British grey seals and 13.6 ppm in northern fur seals. Another report found kidney cadmium levels in common seals from the British Islands to be much lower, 0.1 to 1.9 ppm.[32] No detrimental effects have been associated with these levels of cadmium in marine mammals. Close cooperation will be required between toxicologists and pathologists to identify possible toxicity of heavy metals and to determine effects vs. residue levels. Heavy metals (cadmium, lead, and mercury) do not concentrate in the fetus or significantly cross the placenta in seals,[32] as do the fat soluble organochlorine residues.

Lead has been reported in low concentration in harbor seals. The mean level was 2.31 ppm in the liver of 15 harbor seals, with an average 3.5 ppm in bone.[32] Neither lead nor cadmium appeared to increase in the tissues with age, suggesting that bioconcentration is not significant in these animals at the top of the food chain. Mercury did tend to increase slowly with increasing age in the survey of harbor seals.[32]

As with other pollutants, acute heavy metal poisoning has not been described in sea mammals, to our knowledge. We have seen one case possibly representing lead toxicity in which there was a mild to moderate multifocal acute renal tubular necrosis. The brain also had areas of gliosis, satellitosis, and early neuronal degeneration.

III. METHODOLOGY

At the time of necropsy examination, liver tissue was collected for laboratory analysis. The tissue was wrapped in aluminum foil and frozen until analysis could be performed.

Ten grams of liver tissue was homogenized into a fine paste in a high speed Waring Blendor.® The liver homogenate was then extracted with acetonitrile. This effectively extracted 99% of the chlorinated hydrocarbons from the tissue paste. The acetonitrile solution was then filtered and further cleaned up by passing through a Florisil column. The acetonitrile solution containing the chlorinated hydrocarbons was then extracted with hexane, which recovered about 90% of the chlorinated hydrocarbons. The hexane solvent was then concentrated by evaporation, and run through a Tracor 222 gas chromatograph equipped with electron capture detector. The carrier gas was nitrogen (20 mℓ/min through the column; 60 mℓ/min through the purge). The column was 180 cm long, 6 mm outside diameter, and 2 mm inside diameter. The column was packed with 1.5% SP-2250/1.95% SP-2401 on 100/120 Supelcoport. Injector, column, and detector temperatures were 235°C, 200°C, and 285°C, respectively. The resultant peaks were analyzed and calculated by a 3390A Hewlett-Packard Integrator. Tissue extracts were analyzed for total chlorinated hydrocarbons—DDT, DDE, DDD, and PCB (Arochlor 1242, 1254, 1260).

IV. OBSERVATIONS

Tables 1 and 2 reference many of the world-wide reports of average levels of PCB's, total DDT (DDT, TDE, DDE) and Dieldrin in various species of pinnipeds and cetaceans. These animals were usually harvested or otherwise collected while in a relatively good nutritional state, and therefore adequate blubber layers were available to sample and analyze.

As mentioned previously, Table 1 indicates that pinnipeds collected from industrialized areas such as southern California and the Baltic Sea have higher residue levels in the tissue. Also, migration patterns could partially account for the high levels of PCB and DDT in the fat of California sea lions in Oregon, as these animals migrate from southern California.

The results of these toxicologic surveys are difficult to evaluate, because the number of individual animals of each species is usually limited, and the evaluation of the health status of the animal has not been correlated.

As expected, Table 2 shows that the larger and more pelagic cetaceans tend to have lower tissue residue levels in their blubber, than the species that are more littoral in their range.

Tables 3 to 9 tabulate our data on PCB and DDT liver levels for stranded pinnipeds and cetaceans along the Los Angeles County, California shore. Most of our animals were found to be quite sick, and in most cases were quite emaciated, hence we routinely sampled liver tissue, as an adequate blubber layer was not present. Our levels tend to be quite high, considering that fat levels are usually 5 to 20 times higher than liver levels, based on data from studies in which both liver and fat levels were analyzed from the same animals.

Our data for seals are limited. The elephant seals examined were usually about 4 to 6 months of age, and the harbor seals were estimated to be 1 to 4 months of age. Harbor seal pups are weaned at a very early age, hence they had opportunity to concentrate PCB and DDT from both maternal milk and from fish.

The fur seal with toxoplasmosis was a 12-year-old animal, tagged as a pup in the Pribilof Islands. Her liver residue levels were low as compared to our adult female sea lions (Table 8), possibly due to her arrival in the southern California area recently before death.

Table 4 indicates that there is transplacental transfer of organochlorines in sea lions, as is seen in humans. These pups were from the long-term captive breeding population at

Table 9
ORGANOCHLORINE RESIDUES (PPM) IN CETACEANS[a]

Animal #	Year	Necropsy findings	PCB	DDT	TDE	DDE	Total DDT
33396	1971	Bacterial hepatitis	3.8	1.2	3.8	64.0	69.0
54391	1974	None	4.7	1.2	3.8	62.0	67.0
54677	1974	Parasitic encephalitis	0	0	0.24	0.84	1.08
56738	1974	Parasitic encephalitis	11.0	3.2	4.1	91.0	98.2
69182	1976	Parasitic encephalitis	19.0	0	1.8	20.0	21.8
71058	1977	Parasitic encephalitis	0.5	0	0	2.9	2.9
88565[b]	1980	Trauma	0	0	0	0.1	0.1
91329[c]	1981	Vascular collapse	0	0	0	0	0
91883[d]	1981	Pulmonary hemorrhage	0	0	0.1	0.12	0.22
92048[e]	1981	None	0	0	0.085	0.058	0.143
		Liver level — wet wt. $\overline{X}$	3.9	0.6	1.4	24.0	26.0

[a] Adult common dolphins, unless otherwise shown.
[b] Juvenile blue whale.
[c] Sub-adult pigmy sperm whale.
[d] Pacific white-sided dolphin.
[e] Northern right whale dolphin.

Marineland. The dams from this colony have considerably lower residues than feral animals, because they are fed a fish diet which has minimal contamination. The tissue residue levels in feral fetuses are reported to be considerably higher.

Tables 5 to 8 show our data for PCB and DDT residue levels in necropsied California sea lions, by age, sex, and date, as well as pathologic findings. Adult female sea lions had an average PCB and DDT burden that was lower than the adult males or subadult animals. Perhaps this was because of maternal transfer of the residues to the fetus by way of the placenta and milk.

Almost all of the animals were emaciated, although this was listed only when the condition was particularly severe and no other significant pathology was found. It is not possible to predict residue levels, based on the age or condition of the animal, as there is so much variation within each group. However, our average levels are significantly higher than reported elsewhere, for pinnipeds. Possibly, because of the poor condition or emaciation of the animals in our groups, there had been some mobilization of organochlorine residues from the blubber during catabolism, resulting in elevated liver levels. However, it is more likely that these values are simply a reflection of a greater level of contamination of these animals in the southern California coastal area.

An interesting observation is that one individual animal may have high liver residues, while another animal of the same age, sex, and year of death may have very low levels. It would be helpful to know the migration pattern of individual animals. Many animals migrate north and south from the Channel Island breeding areas off the southern California coast, while others remain year round, presumably in more polluted waters with access to more highly contaminated fish.

In southern California, the coastal water around Los Angeles is most contaminated by sludge from sewage outfalls that dump into the ocean 5 to 7 miles into Santa Monica Bay, 1 to 2 miles off Palos Verdes Peninsula, and into San Pedro Bay (Figure 1).[3,45] In this zone, the fish have higher PCB and DDT residue levels in their tissues, compared to other coastal areas.[45] The most highly contaminated fish are found off the Palos Verdes Peninsula.[45] There is a colony of about 50 sea lions on the point of the peninsula (Figure 2). They spend about half the year there and hence are likely to consume fish with a higher tissue organochlorine burden. While we have no confirmed data on this particular group of animals, we presume

FIGURE 1. Map of Los Angeles County, California shoreline. The shaded area off the southern coast shows the area of highest relative concentration of PCB and DDT in the sediment. The dark dots represent the offshore sewage outfalls. Note the location of Santa Monica Bay (1), Palos Verdes Peninsula (2), and San Pedro Bay (3). (Adapted from Howard, E. B., Esra, G. N., and Young, D., in *Animals as Monitors of Environmental Pollutants*, Nielsen, S. W., Migaki, G., and Scarpelli, D. G., Eds., National Academy of Sciences, Washington, D.C., 1979, 290. With permission.)

that they will be found to have higher tissue residues than the more gregarious sea lions that seem to commute from the offshore islands, away from the most heavily polluted feeding grounds.

Table 9 shows that of the cetaceans we have studied, the common dolphin had higher organochlorine levels than other species which are more pelagic and wide ranging. The juvenile blue whale had been impaled on the bottom of a ship off Central America, in more "clean waters", and had liver level of only 0.1 ppm of DDT.

The majority (about 90%) of the total DDT found in pinnipeds and cetaceans was in the form of DDE. The residue pool of DDT compounds in the bottom sediment along the Los Angeles coastal waters is quite abundant and persistant, and will apparently remain for an indeterminant period of time. We have continued to find levels of undegraded DDT in the livers of marine mammals as recently as 1981, over a decade since the cessation of use of this compound. Actually, the buildup of this material in the off shore drainage pipes was so significant over the years, that there are still detectable levels present in the effluent being dumped into the offshore waters along this coast. We have demonstrated that the uptake and concentration of these compounds in bottom feeding fish can result in acute toxicity in other species of fish eating animals,[3] and we do not know if there is present insidious toxicity in marine mammals. We feel this possibility is overdue for study.

FIGURE 2. Colony of approximately 50 sea lions on the point of Palos Verdes Peninsula.

REFERENCES

1. **Rosewell, K. T., Muir, D. C. G., and Baker, B. E.,** Organochlorine residues in harp seals (*Phagophilus groenlandicus*) tissues. Gulf of St. Lawrence, 1971, 1973, *Pest. Monitoring J.*, 12, 189, 1979.
2. **Kutz, F. W., Strassman, S. C., and Sperling, J. F.,** Survey of selected organochlorine pesticides in the general population of the United States: fiscal years 1970—1975, in *Health Effects of Halogenated Hydrocarbons,* Nicholson, W. J. and Moore, J. A., Eds., New York Academy of Sciences, New York, 1979, 60.
3. **Howard, E. B., Esra, G. N., and Young, D.,** Acute foodborne pesticide toxicity in comorants (*Phalacrocorax* sp.) and seagulls (*Larus californicus*), in *Animals as Monitors of Environmental Pollutants,* Nielsen, S. W., Migaki, G. and Scarpelli, D. G., Eds., National Academy of Sciences, Washington, D. C., 1979, 290.

4. **Matthews, H. B. and Kato, G.,** The metabolism and disposition of halogenated aromatics, in *Health Effects of Halogenated Hydrocarbons,* Nicholson, W. J. and Moore, J. A., Eds., New York Academy of Sciences, New York, 1979, 131.
5. **Anderson, R. S.,** Metabolism of an environmental carcinogen by invertebrate animals, in *Animal Models of Comparative and Developmental Aspects of Immunity and Disease,* Gershwin, M. E. and Cooper, E. L., Eds., Pergamon Press, New York, 1978, 25.
6. **Zechmeister, L. and Koe, B. K.,** The isolation of carcinogenic and other polycyclic aromatic hydrocarbons from barnacles, *Arch. Biochem. Biophys.,* 35, 1, 1952.
7. **Hu, T. C. L., Gossett, R., and Young, D.,** Mass spectrometry confirmation of PCB and DDT analyses of fish, *Southern California Coastal Water Research Project Biennial Report,* Long Beach, Calif., 1980, 131.
8. **Dustman, E. H. and Stickel, L. F.,** The occurrence and significance of pesticide residues in wild animals, in *Biological Effects of Pesticides in Mammalian Systems,* Kraybill, H. F., Ed., New York Academy of Sciences, New York, 1969, 162.
9. **LeBoeuf, B. J. and Bonnell, M. L.,** DDT in California sea lions, *Nature (London),* 234, 108, 1971.
10. **Koller, L. D.,** Effects of environmental contaminants on the immune system, *Adv. Vet. Sci. Comp. Med.,* 23, 267, 1979.
11. **Cavagnolo, R. Z.,** The immunity of marine mammals, *Develop. Comp. Immunol.,* 3, 245, 1979.
12. **Anas, R. E. and Wilson, A. J.,** Organochlorine residues in fur seals, *Pest. Monitoring J.,* 3, 198, 1970.
13. **Kurtz, D. A. and Kim, K. C.,** Chlorinated hydrocarbon and PCB residues in tissues and lice of northern fur seals, 1972, *Pest. Monitoring J.,* 10, 79, 1976.
14. **Robinson, J. and Hunter, C. G.,** Organochlorine insecticides: concentrations in human blood and adipose tissue, *Arch. Environ. Health,* 13, 558, 1966.
15. **Durham, W. F.,** Body burden of pesticides in man, in *Biological Effects of Pesticides in Mammalian Systems,* Kraybill, H. F., Ed., New York Academy of Sciences, New York, 1969, 183.
16. **Addison, R. F. and Brodie, P. F.,** Organochlorine residues in maternal blubber, milk, and pup blubber from grey seals (*Halichoercus grypus*) from Sable Island, Nova Scotia, *J. Fisheries Res. Bd. (Canada),* 34, 937, 1977.
17. **Helle, E., Olsson, M., and Jensen, S.,** PCB levels correlated with pathological changes in seal uteri, *Ambio,* 5, 261, 1976.
18. **Zavon, M. R., Tye, R., and Latorre, L.,** Chlorinated hydrocarbon insecticide content of the neonate, in *Biological Effects of Pesticides in Mammalian Systems,* Kraybill, H. F., Ed., New York Academy of Sciences, New York, 1969, 196.
19. **DeLong, R., Gilmartin, W. G., and Simpson, J.,** Premature births in California sea lions: association with organochlorine pollutant residue levels, *Science,* 181, 1168, 1973.
20. **Fry, D. M. and Toone, C. K.,** DDT-induced feminization of gull embryos, *Science,* 213, 922, 1981.
21. **Arly, J. A., Struck, R. F., and James, R.,** Estrogenic activities of chlorinated hydrocarbons, *J. Toxicol. Environ. Health,* 4, 325, 1978.
22. **Wasserman, M., Wasserman, D., Gershon, Z., and Zellermayer, L.,** Effects of organochlorine insecticide on body defense systems, in *Biological Effects of Pesticides in Mammalian Systems,* Kraybill, H. F., Ed., New York Academy of Sciences, New York, 1969, 393.
23. **Peakall, D. B.,** Pesticide-induced enzyme breakdown of steroids in birds, *Nature (London),* 216, 505, 1967.
24. **Allen, J. R., Lambrecht, L. K., and Barsotti, D. A.,** Effects of polybrominated biphenyls in non-human primates, *J.A.V.M.A.,* 173, 1485, 1978.
25. **Anderson, D. W., Hickey, J. J., Risebrough, R. W., Hughes, D. F. and Christensen, R. E.,** Significance of chlorinated hydrocarbon residues to breeding pelicans and cormorants, *Can. Field-Naturalist,* 83, 91, 1969.
26. **Phillips, W. E. J.,** DDT and the metabolism of vitamin A and carotene in the rat, *Can. J. Biochem. Physiol.,* 41, 1793, 1963.
27. **Tinsley, I. J.,** Nutritional interactions in dieldrin toxicity, *Agric. Food Chem.,* 14, 563, 1966.
28. **McConnell, E. E. and Moore, J. A.,** Toxicopathology characteristics of the halogenated aromatics, in *Health Effects of Halogenated Aromatic Hydrocarbons,* Nicholson, W. J. and Moore, J. A., Eds., New York Academy of Sciences, New York, 1979, 138.
29. **Smith, W. E. and Smith, A. M.,** *Minimata,* Holt, Rinehart, and Winston, New York, 1975, 182.
30. **Koller, L. D.,** Public health risks of environmental contaminants: heavy metals and industrial chemicals, *J.A.V.M.A.,* 176, 525, 1980.
31. **Buhler, D. R., Claeys, R. R., and Mate, B. R.,** Heavy metal and chlorinated hydrocarbon residues in California sea lions *(Zalophus californianus), J. Fisheries Res. Bd. (Canada),* 32, 2391, 1975.
32. **Roberts, T. M., Heppleson, P. B., and Roberts, R. D.,** Distribution of heavy metals in tissues of the common seal, *Marine Pollut. Bull.,* 7, 194, 1976.

33. **Magos, L. and Webb, M.,** The interactions of selenium and mercury, *Critical Reviews in Toxicology,* Vol. 8, Goldberg, L., Ed., CRC Press, Boca Raton, Fla., 1980, 1.
34. **Holden, A. V. and Marsden, K.,** Organochlorine pesticides in seals and porpoises, *Nature (London),* 216, 1274, 1967.
35. **Clausen, J., Braestrup, L., and Berg, O.,** The content of polychlorinated hydrocarbons in Arctic mammals, *Bull. Environ. Contam. Toxicol.,* 12, 529, 1974.
36. **Addison, R. F., Kerr, S. S., Dale, J., and Sergeant, D. E.,** Variation of organochlorine residue levels with age in Gulf of St. Lawrence harp seals *(Pagophylus groenlandicus), J. Fisheries Res. Bd. (Canada),* 30, 595, 1973.
37. **Takei, G. H. and Leong, G. H.,** Macro-analytical methods used to analyze tissues of the Hawaiian monk seal, *Monachus schauinslandi,* for organochlorine pesticides, polychlorobiphenyls, and pentachlorophenol, *Bull. Environ. Contam. Toxicol.,* 27, 489, 1981.
38. **Wolman, A. A. and Wilson, A. J.,** Occurrence of pesticides in whales, *Pest. Monitoring J.,* 4, 8, 1970.
39. **Taruski, A. G., Olney, C. E., and Winn, H. E.,** Chlorinated hydrocarbons in cetaceans, *J. Fisheries Res. Bd. (Canada),* 32, 2205, 1975.
40. **Saschenbrecker, P. W.,** Levels of DDT and PCB compounds in north Atlantic fin-back whales, *Can. J. Comp. Med.,* 37, 203, 1973.
41. **Gaskin, D. E., Smith, G. J. D., Arnold, P. W., Louisy, M. V., Frank, R. Holdrinet, M., and McWade, J. W.,** Mercury, DDT, dieldrin, and PCB in two species of Odontoceti (Cetacea) from St. Lucia, Lesser Antilles, *J. Fisheries Res. Bd. (Canada),* 31, 1235, 1974.
42. **Lehman, J. W. and Peterle, T. J.,** DDT in cetacea, *Invest. Cetacea,* 3, 349, 1971.
43. **Addison, R. F. and Brodie, P. F.,** Occurrence of DDT residues in beluga whales (*Delphinapterus leucas*) from the Mackenzie Delta, N.W.T., *J. Fisheries Res. Bd. (Canada),* 30, 1733, 1973.
44. **Koeman, J. H., Peeters, W. H. M., Smith, C. J., Tjioe, P. S., and DeGoeij, J. J. M.,** Persistent chemicals in marine mammals, *T. N. O. Nieuws,* 27, 570, 1972.
45. **Sherwood, M. J.,** The fin erosion syndrome, in *Southern California Coastal Water Research Project Annual Report,* El Segundo, Ca., 1978, 203.

Chapter 5

NEOPLASMS IN MARINE MAMMALS

Edwin B. Howard, James O. Britt, Jr., and John G. Simpson

TABLE OF CONTENTS

I. NEOPLASMS

Neoplastic disease, in past years, has been considered an uncommon or rare event in marine mammals. It was even speculated that they may have some inherent resistance to the development of neoplasms. However, in more recent years, with more careful observation of large numbers of stranded and beached animals, the cumulative number of observed neoplasms in these marine animals is constantly growing.

Many of the earlier case reports on marine mammal neoplasms were derived from studies of captive zoo animals. In many of the earlier reports, the histopathologic terminology is vague or inconsistent with the terminology used today, making a clear understanding of the tumor type and origin difficult. The reports of neoplasms of marine mammals have come mostly as a result of examination of whales caught commercially by whalers, zoos, oceanariums, examination of dead beached animals, and from institutions which maintained and studied these animals for research purposes. Therefore, the species represented may be biased to over-represent whales, dolphins, and sea lions. Our study is also biased in favor of recognition of tumors in the California Sea Lion, because this animal accounted for 2/3 of those examined as part of the stranded animal surveillance program.

A comprehensive literature review of diseases in wild animals, completed in 1955,[1] included only three reports of tumors in pinnipeds and five reports describing tumors in cetaceans. The pinniped tumors included an adenocarcinoma and myosarcoma of the mammary gland, with metastasis to the lung,[2] a hypernephroma of the adrenal gland,[3] and an unidentified tumor of the optic nerve.[4] Those reports describing tumors in cetaceans included an unidentified tumor of the stomach of a dolphin,[5] which may have been an inflammatory lesion associated with parasitism,[6] a ganglioneuroma in the mediastinum of a whale,[7] three granulosa cell tumors and a cystadenoma of the ovary of whales, and a papilloma of the tongue of a whale.[8] Stolk described five cases of fibromas in whales, an ovarian carcinoma in a fin whale, a case of Hodgkins disease in a fin whale, a hemangioma of the liver of a whale, and a liver adenoma in a dolphin.[9-12]

Simpson and Gardner[13] described a fibroma of the neck and an adenocarcinoma with involvement of the skin and lung, both in California Sea Lions. The adenocarcinoma was thought to be primary in the skin, probably of sweat gland origin, and metastasized to the lung.[14]

Ridgway compiled a tabulation of marine mammal diseases which included 12 reports of neoplasms in cetacea and seven reports of tumors in pinnipeds.[15] Of those cetacean tumors tabulated, a benign melanotic tumor of the lip of a whale,[16] a benign lipoma in a fin whale,[17] and a lipoma of the brain in a humpback whale[18] have not previously been mentioned in this chapter. Landy presented a literature review of tumors in marine mammals up to 1978, wherein he described 31 tumor reports in cetacea and 40 reports of tumors in pinnipeds.[19] Based on this summary review of marine mammal neoplasms, it would appear that neoplasms are somewhat more common in pinnipeds than in cetaceans, but this may be biased because of a greater sampling of pinnipeds. Malignant neoplasms appear to be more frequent in pinnipeds, with benign tumors more often described in cetaceans.

Because of the common practice among marine mammal investigators of sending tissue samples or specimens from interesting cases to several individuals or institutions, it is possible for the same case to be tabulated or described more than once. If possible, a person should try to ascertain if this has happened before publishing material on a case in which the necropsy examination was performed by someone else. Also, the use of personal communication of case material results in duplication of reports, for the same stated reason. As an example, in the above mentioned review of tumors in marine mammals, the table on neoplasia in pinnipeds contains at least two, and possibly more duplications. From information available, References 13, 27, and 41 are dealing with the same tumor; and it seems that References

17 and 26 are also referring to the same carcinoma of the skin of the neck with spread to the lungs. Because of tabulation of data such as this, accurate figures concerning frequency of tumor occurrence may be difficult to obtain. Effort should be made in reviewing the literature and compiling data from personal communication to avoid duplicate reporting of the same animal.

As can be seen from Table 1, neoplasms in cetaceans appear to be rather infrequent in occurrence, especially when taking into consideration the many whales examined by scientists accompanying whaling ships and at whale processing stations. In one report on the examination of 2000 whales, only two were found to have tumors.[16] One was a benign fibromyoma of the uterus of a 19-year-old female sperm whale, and the second was a small benign melanotic lesion, possibly a hamartoma, on the lip of a 48-year-old female sei whale. In a review of killer whales captured for display in oceanariums, most of the animals recorded subsequently died, and no neoplasms were recorded in this small group of animals.[40]

The majority of tumors of cetaceans are benign, with only five of 36 (14%) neoplasms histologically described as being of a malignant type. Considering anatomic sites, 12 of the 36 tumors of cetaceans reported in the literature involved the genital tract, and considering tumor types, 17 of the 36 tumors or almost 50% of the total, were histologically classified as fibromas or lipomas.

Tumors of pinnipeds, on the other hand, have been more frequently reported, with a total of 58 neoplasms found in a literature survey (Table 2). The greater number of tumors in pinnipeds is undoubtedly influenced by the fact that much greater numbers of these animals have been examined as stranded feral animals, as well as being more adaptable to zoo and oceanarium environments around the world.

It does become readily apparent on perusing the pinniped tumors reported in the literature (see Table 2), that they are much more likely to be malignant, with 34 of 58 or nearly 60% of them being histologically classified as malignant. It is interesting that tumors of the genital system occur with similar frequency in cetaceans (33%) and pinnipeds (28%). Neoplastic disease of the hematopoietic system appears to be over-representated in pinnipeds, with 10 of 54 cases being either lymphoid malignancy (9) or myeloid leukemia (1).

A compilation of tumors found in our survey of marine mammals has revealed 39 neoplasms, of which four were in cetaceans and 35 in pinnipeds. Our case material was considerably biased, with over 2/3 of the animals studied being pinnipeds. A summarization of our tumor cases is presented in Table 3, and representative examples of the neoplasms detected will be demonstrated for the various organ systems. The apparent sexual bias, most cases being in females, is because the male animals on off-shore islands, seldom are seen stranded on beaches.

II. INTEGUMENT

Only three neoplasms were identified in the skin. One of these was a highly anaplastic and undifferentiated adenocarcinoma of the skin of the neck of a California sea lion (Figure 1).[13] The tumor was highly infiltrative through the subcutis and underlying muscle tissue, and had metastasized to the lung (Figure 2). The tumor cells were large and pleomorphic and had abundant clear cytoplasm. The tumor was believed to have arisen from sweat glands of the skin of the neck of this animal.

The second was an excision biopsy of a chronic ulcerated, nonhealing skin lesion from a Pacific white-sided dolphin (Figure 3). Skin sections from this lesion revealed a moderately well differentiated squamous cell carcinoma, with multiple finger-like projections of the tumor cells extending down into the dermis under the area of epidermal ulceration (Figure 4), and extending laterally under the intact epidermis, at the edges of the lesion. There was moderate pleomorphism of the tumor cells (Figure 5), with a moderate degree of mitotic

Table 1
TUMORS OF CETACEANS

Animal	Age	Sex	Tumor site	Tumor type	Ref.
Common porpoise	Adult	M	Penis	Papilloma	19
Bottle-nose dolphin	Adult	M	Testicle	—	20
	—	M	Liver	Adenoma	10
	10—15	M	Pancreas	Carcinoma	19
	—	F	Lung, liver	Reticuloendotheliosis	19
	20+	F	Spleen	Lymphosarcoma	19
	—	—	Kidney	Adenoma	41
	—	—	Adrenal	—	20
	10—15	M	Thyroid	Adenoma	19
Killer whale	6—7	M	Penis	Papilloma	19
Pilot whale	31	F	Uterus	Leiomyoma	21
Sperm whale	19	F	Uterus	Fibromyoma	16
	—	—	Jaw	Fibroma	10
	—	—	Liver	Hemangioma	11
	—	M	Skin	Fibroma	11
Beaked whale	Adult	F	Vagina	Fibromas (2)	22
Blue whale	—	F	Uterus	Fibromyoma	19
	—	F	Ovary	Cystadenoma	8
	—	F	Ovary	Granulosa cell tumor	8
	—	—	Mediastinum	Ganglioneuroma	7
	—	—	Tongue	Papilloma	8
	—	F	Stomach serosa	Lipoma	18
	—	—	Pleura	Fibroma	20
	—	F	Liver surface	Lipoma	18
	—	—	Intestine serosa	Lipoma	18
Fin whale	—	F	Ovary	Granulosa cell tumor	8
	—	F	Ovary	Granulosa cell tumor	8
	—	F	Ovary	Carcinoma	9
	Adult	—	Cerebellum	Neurofibroma	23
	—	F	Tongue	Fibroma	10
	—	M	Multicentric	Hodgkin's disease	12
	—	—	Pleura	Fibroma	20
	—	M	Subcutis	Fibroma	10
	—	M	Skin	Fibroma	11
	—	—	Dorsal muscle	Lipoma	18
Bowhead whale	Immature	M	Liver	Lipoma	60

Note: — = not given.

Table 2
TUMORS OF PINNIPEDS

Animal	Age	Sex	Tumor site	Tumor type	Ref.
California sea lion	—	F	Mammary gland	Adenocarcinoma & myosarcoma	2
	—	—	Adrenal	Hypernephroma	3
	—	—	Optic nerve	—	4
	—	—	Neck	Fibroma	13
	—	—	Skin adnexa	Adenocarcinoma	13
	3	M	Multicentric	Lymphosarcoma	19
	—	F	Pancreatic ducts	Adenocarcinoma	19
	—	F	Pituitary	Adenoma	19
	—	F	Pancreas, pituitary, adrenal	Adenomas (multiple)	19

Table 2 (continued)
TUMORS OF PINNIPEDS

Animal	Age	Sex	Tumor site	Tumor type	Ref.
	1 month	F	Pituitary	Adenoma	19
	2.2	F	Adrenal	Adenoma	19
	22	F	Mammary gland	Fibrosarcoma	19
	Adult	—	Adrenal	Adenocarcinoma	19
	—	F	Uterus	Leiomyoma	19
	—	—	Kidney	Transit. cell carcinoma	19
	—	F	Abdomen	Clear cell carcinoma	19
	—	F	Pharynx, esophagus	Lipoma	19
	—	—	Pancreas	—	19
	—	F	Skin	Squamous cell carcinoma	19
	—	M	Perineum	Squamous cell carcinoma	19
	—	M	Eye	Malignant melanoma	24
	—	—	—	Endocrine carcinoma	25
	Adult	F	Uterus (possibly)	Adenocarcinoma	26
	Adult	M	Undetermined	Adenocarcinoma	26
	—	—	Pharynx	Squamous cell carcinoma	27
	16	F	Pancreatic ducts	Adenoma	28
	16 months	—	Kidney	Nephroblastoma	29
	Adult	F	—	Squamous cell carcinoma	29
	—	—	Abdominal cavity	Mesenchymoma	30
	Young	—	Multicentric	Lymphosarcoma	32
Harbor seal	2.5 month	M	Multicentric	Lymphosarcoma	31
	1.5 month	M	Multicentric	Lymphosarcoma	31
	Young	—	Multicentric	Lymphosarcoma	25
	Young	—	Multicentric	Lymphosarcoma	32
Northern fur seal	Neonate	—	Multicentric	Lymphosarcoma	33
	—	M	Prepuce & penis	Squamous cell carcinoma	27
	2 weeks	—	Kidney	Fibrosarcoma	34
	—	—	Liver	Lipoma	35
	Adult	F	Ovary	Granulosa cell tumor	27
Southern elephant seal	Adult	F	Ovary	Granulosa cell tumor	36
	—	M	Eye	—	37
	—	M	Subcutis	—	19
	—	F	Adrenal	Adenoma	19
Gray seal	44	F	Cervix & uterus	Leiomyomas (multiple)	38
			Uterus	Squamous cell carcinoma	38
Harp seal	Adult	—	Multicentric	Lymphosarcoma	19
Alaskan fur seal	—	F	Vagina & cervix	Squamous cell carcinoma	27
Cape fur seal	—	F	Ovary	Granulosa cell tumor	19
Seal	—	—	Multicentric	Lymphosarcoma	35
Walrus	4	F	Multicentric	Myeloid leukemia	35
Sea otter	Adult	F	Uterus	Leiomyoma	39
	Adult	F	Uterus	Leiomyoma	39
	Adult	F	Uterus	Leiomyoma	54
			Adrenal	Pheochromocytoma	54
			Liver	Bile duct carcinoma	54
Spot-neck otter	13 months	F	Ovary	Teratoma (malignant)	55

Note: — = not given.

Table 3
NEOPLASMS OF MARINE MAMMALS (SOUTHERN CALIFORNIA STUDY)

Animal	Age	Sex	Tumor site	Tumor type
California sea lion	Aged	F	Lung	Epidermoid carcinoma
	Adult	F	Lung	Epidermoid carcinoma
	Adult	M	Lung	Epidermoid carcinoma
	Adult	F	Lung	Epidermoid carcinoma
	Adult	M	Lung	Adenocarcinoma
	Adult	F	Liver	Bile duct adenocarcinoma
	Aged	F	Liver	Bile duct adenocarcinoma
	Adult	F	Liver	Bile duct adenocarcinoma
	Adult	M	Liver	Bile duct adenocarcinoma
	Adult	F	Liver	Bile duct adenocarcinoma
	Adult	F	Liver	Bile duct adenocarcinoma
	Adult	F	Liver	Bile duct adenocarcinoma
	Adult	F	Uterus	Leiomyoma
	Aged	F	Uterus	Leiomyoma
	Adult	F	Uterus	Leiomyoma
	Adult	F	Uterus	Adenocarcinoma
	Adult	F	Uterus	Adenocarcinoma
	Adult	F	Ovary	Adenocarcinoma
	Adult	F	Ovary	Adenocarcinoma
	Adult	F	Ovary	Adenocarcinoma
	Adult	F	Ovary	Adenocarcinoma
	Adult	F	Ovary	Adenoma
	Adult	F	Ovary	Granulosa cell tumor
	Adult	F	Cervix & vagina	Squamous cell carcinoma
	Adult	F	Vagina	Papilloma
	Adult	F	Vagina	Hemangioma
	Adult	M	Prostate	Adenocarcinoma
	Adult	F	Mammary gland	Lipoma (infiltrating)
	Adult	F	Kidney	Adenocarcinoma
	Adult	F	Kidney	Adenocarcinoma
	Adult	F	Urinary bladder	Transitional cell carcinoma
	Adult	F	Urinary bladder	Papilloma
	Adult	M	Kidney & adjacent	Neuroendocrine carcinoma
	—	—	Skin (met. to lung)	Adenocarcinoma[13]
	Adult	F	Skin	Leiomyoma
Pacific white-sided dolphin	Adult	M	Skin	Squamous cell carcinoma
	Adult	M	Lymph nodes, spleen	Lymphosarcoma
	Adult	M	Generalized	Eosinophilic leukemia
	Adult	F	Mouth	Gingival fibroma

activity. Rare areas of keratinization of tumor cells were found. Treatment of the lesion appeared to be curative, with no further evidence of neoplasia. Cryosurgery was the method of treatment of the lesion.

The third skin tumor was a firm nodule in the dermis of a sea lion, which on histologic evaluation was found to be a benign leiomyoma. The tumor elevating the epidermis, expanding and compressing the subcutaneous fat, and the tumor cells were quite well differentiated (Figure 6) with long sausage shaped nuclei and densely packed elongated cells. The origin is presumed to be either smooth muscle associated with hair follicles or muscular artery in the dermis.

III. LUNG

In the group of animals we have studied, there have been five neoplasms that we have concluded were primary in the lung. Of the five, the histologic features of four were typical of epidermoid or squamous cell carcinoma, indicating probable origin from main stem or large bronchial lining epithelium, and one was adenocarcinoma of bronchial origin.

Those lung tumors which were epidermoid or squamous cell carcinoma were either moderately differentiated tumors with sheets and islands of cells with minimal keratinization (Figure 7), or the tumors were well differentiated epidermoid carcinomas with keratin pearls (Figure 8). Some areas of a single tumor would show more differentiation than other areas. The tumors randomly infiltrated through the lung parenchyma, often filling bronchioles and alveolar sacs, and lymphatic invasion was detectable. Metastases were present in all cases of epidermoid carcinoma, with involvement of bronchial nodes, and distant metastases to adrenals, spleen, and liver were often found.

The other primary lung tumor was a poorly differentiated adenocarcinoma, which showed tendency toward acinar development (Figure 9). However, in both the primary lesion and the metastatic foci in lymph nodes, there was extensive squamous metaplasia of the tumor cells (Figure 10). There were no keratin pearls or bridges found in the areas of squamous metaplasia.

IV. LIVER

The liver was a frequently observed primary site of neoplastic disease, with seven California sea lions presented with bile duct adenocarcinoma. The pattern of distribution of tumor throughout the liver was variable. In most cases there were large firm white infiltrative masses, often with multiple nodules widely separated, and there was usually central necrosis in the larger tumor masses. Dissemination was always present, with lymph node, spleen and lung metastases. One rather unusual case was seen in which there were numerous widely scattered small foci of tumor throughout the liver, with no larger discreet tumor mass detected (Figure 11). Careful examination of all other organs failed to reveal any other possible source of the neoplasm, and metastases were found in the lymph nodes adjacent to the liver.

It is interesting that on the histologic examination of the bile duct tumor seen in Figure 11, it was found to be quite well differentiated, forming a ductal or acinar pattern of cuboidal appearing cells, surrounding varying sized pools of mucinous material (Figure 12). Using PAS stain, it was found that the trapped material was indeed mucin (Figure 13), and that the metastatic foci in the nodes and lung maintained this same histologic pattern, and with mucin production. This pattern of mucin production by intrahepatic bile duct adenocarcinoma was unique to this tumor of this series of bile duct tumors, although it is a common feature of similar tumors in other species.

The bile duct adenocarcinomas found in the other sea lions were much less differentiated, and often showed a solid pattern, with sheets of tumor cells expanding on and infiltrating the adjacent liver parenchyma (Figures 14 and 15). This tumor pattern usually contained minimal glandular differentiation, and often contained areas of squamous metaplasia, in both the primary sites in the liver, and in the metastatic foci (Figure 16). One of the bile duct adenocarcinomas was characterized by large tumor cells with clear cytoplasm, forming sheets of infiltrating tumor without significant ductal differentiation (Figure 17).

V. GENITAL SYSTEM

Considering the genital system in its entirety, this system accounted for the largest single number of tumors in our series, with a total of 16 neoplasms, eight of which were malignant and eight benign.

Smooth muscle tumors of the uterus has been a frequently recognized tumor in the past reports, and we have recognized three such tumors, leiomyomas, in our series of animals. One tumor was a large mass, about 4×5×5 in. in size and was located in the body of the uterus at the bifurcation (Figure 18A). The other two were in the horn of the uterus, and the smallest was a nodular mass attached to the surface of the uterine horn (Figure 18B). All were well differentiated tumors composed of densely packed smooth muscle cells which contained elongated, rounded-end nuclei, and an absence of mitotic cells (Figure 19). Use of Van Gieson stain revealed sparse collagen fibers running among the smooth muscle cells, but collagen was an insignificant feature of the tumors.

Adenocarcinoma of the uterus was found in two animals, and there were widespread abdominal metastases in both cases. There was considerable necrosis within the center of the main tumor mass in the uterus. Examination of sections from the uterine wall demonstrated continuity with the uterine mucosa, with invasion of tumor into and through the muscular wall to the serosal surface. One uterine adenocarcinoma was well differentiated, with mucin filled glandular structures formed by the tumor cells invading the wall of the uterus (Figures 20 and 21). The other adenocarcinoma arising in the uterine mucosa was less well differentiated, and consisted of sheets of tumor cells with minimal glandular pattern, and evidence of squamous metaplasia in both the primary sites and in the abdominal metastases (Figures 22 and 23).

Ovarian tumors were detected in six California sea lions. Four of the tumors were adenocarcinoma, one was an adenoma, and one was a granulosa cell tumor. The ovarian adenocarcinomas were masses involving the ovary and often the oviduct adjacent to the ovary. These tumors were fairly well differentiated (Figures 24 and 25) and in addition to nodal and hepatic metastasis, there was seeding of the peritoneal surfaces by tumor cells which developed a papillary to glandular pattern (Figure 26). There was also a moderate lymphoid response associated with the primary and metastatic foci of these tumors.

The single ovarian adenoma was found incidentally in an adult female which stranded and died. The right ovary was 4×5×5 cm in size, white, and somewhat irregular in surface appearance. It was soft and friable on cutting. It was a well differentiated tumor, composed of cuboidal to low columnar epithelial cells forming well defined acinar structures (Figure 27). There was no evidence of invasion or metastasis. The uterus was normal.

The single case of granulosa cell tumor in a California sea lion was found as a large mass, 6×6×7 cm diameter, largely obliterating the left ovary. On cut section there were cystic spaces in the tumor. Histologic examination revealed a neoplasm which had largely replaced the normal ovarian tissue (Figure 28), and the tumor was composed of very loosely arranged hyperchromatic tumor cells aligned on a basement membrane (Figure 29). The tumor cells were moderately pleomorphic, with considerable nuclear atypia and marked variation in nuclear-cytoplasmic ratio. Although this neoplasm appeared histologically malignant, as is often the case in other mammals, there were no detectable metastases found.

Squamous cell carcinoma of the cervix was found in one California sea lion. It was a moderately well differentiated neoplasm composed of keratinizing squamous epithelial cells (Figure 30), which were destroying the cervical area, and had extensively invaded the vagina and through the cervix into the body of the uterus. In another adult female sea lion that died, there was a large hemorrhagic mass found involving the wall of the vagina, with protrusion into the lumen and partial obstruction of the cavity of the vagina. On tissue sections this was found to be a connective tissue tumor composed of varying sized vascular spaces, some cavernous, (Figure 31) lined by endothelial cells, and with a moderate connective tissue stroma. This was considered to be an example of a well differentiated benign hemangioma, and probably was responsible for some chronic blood loss from this animal.

The vagina was also the site of a benign polyp or papillomatous polypoid structure which was possibly of inflammatory origin in a mature female California sea lion. This was a cauliflower like structure that partially obliterated the lumen of the vagina. The fronds of the papillomatous structure had a surface lining of squamous to low epithelial cells, and these fronds had a prominent fibrovascular supporting structure which was moderately infiltrated by lymphocytes and plasma cells (Figures 32 and 33). On high power microscopic examination, small bodies were seen in these papillomatous structures, which may have been a protozoal organism although this has not yet been proven.

In one adult to aged California sea lion, a mass was found in the pelvic inlet region in a site compatible with the prostate gland. The bladder was not affected and the adjacent colon was not involved. There were metastases from this neoplasm to the sublumbar nodes. Histologic examination indicated this mass was a well differentiated acinar forming adenocarcinoma, (Figure 34) and the acinar structures contained some PAS reactive mucin material. This was an adenocarcinoma and believed to have originated in the prostate gland of this animal. The metastases were also found to be well differentiated acinar forming tumor nodules.

One adult female California sea lion was found stranded on the beach, and was in good condition except for a large superficially ulcerated mass, $10 \times 25 \times 35$ cm in size, involving the right mammary gland, about five times the size of the opposite gland. The surface of the mass contained a 15×20 cm diameter ulcerated area which exuded a milky fluid. The animal was euthanized and a necropsy examination performed. On examination of sections from this mammary mass, it was found to be a lipomatous tumor which was diffusely infiltrating among the tubular and acinar components of the gland (Figure 35). The neoplastic tissue consisted of a mixture of well differentiated, fat filled lipocytes of varying size, and nests of lipocytes which had homogeneous slightly eosinophilic cytoplasm, with only small lipid droplets or no evidence of cytoplasmic lipid (Figure 36). There were no mitotic cells readily seen, and there was no cellular anaplasia, and the lesion was considered to be a benign infiltrating lipoma. There was no evidence of disease other than in the mammary gland.

VI. URINARY TRACT

Four neoplasms were found in adult California sea lions which were primary in the urinary tract. One was a tubular adenocarcinoma arising in the left kidney of an adult female. It had mostly destroyed the kidney, and there were widespread metastases, including abdominal nodes, liver, spleen, and lung. The neoplasm was composed of tumor cells which were attempting to form duct like structures, but the tumor cells were piling up into several layers deep (Figure 37). There was a moderate amount of squamous metaplasia of the tumor cells, which was more prominent in the metastatic sites (Figure 38).

One aged female California sea lion was recently discovered in a terminal state on the beach. At necropsy examination there were multiple firm whitish nodules in both kidneys (Figure 39), and there were numerous metastases in other tissues. Genital organs were not involved. Histologically, this was found to be an anaplastic renal tubular adenocarcinoma, with minimal or poor differentiation of tubular structures (Figure 40). Interestingly, the metastases into other tissues, such as the heart, revealed a much more anaplastic tumor, with marked spindle cell features (Figure 41), with barely detectable suggestions of tubular development. This is the only adenocarcinoma in our series that showed this type of anaplasia, in contrast to other adenocarcinomas which often showed squamous metaplasia.

One transitional cell carcinoma was found in the urinary bladder and surrounding tissue of a California sea lion. This tended to cause extensive local obliteration of urinary bladder wall and adjacent uterus. There were metastases to abdominal nodes, liver and lung in this case. The tumor cells were piling up in multiple layers of squamoid cells with moderate cellular anaplasia (Figure 42), and sheets of squamoid appearing cells were prominent in the metastatic lesions (Figure 43) of the mesenteric nodes, liver and lung. There was moderate bilateral hydroureter and hydronephrosis in this case, due to urinary flow obstruction.

Also found in the urinary bladder of an adult female California sea lion was a papillomatous proliferation of the transitional cells, results in frond-like projections of squamoid epithelial cells with the lumen of the bladder. A portion of the lesion is shown in Figure 44, where the change from normal transitional epithelium to papilloma can be seen. There was a minimal fibrous or vascular stroma in the fronds of tumor cells (Figure 45).

A case of renal adenoma has been described in an Atlantic bottle-nosed dolphin,[41] and this is shown in Figure 46. It was a well differentiated tumor, composed of cuboidal epithelium forming tubules (Figure 47) and was a discreet nodule in one pole of the kidney.

VII. HEMATOPOIETIC

Two cases of neoplastic disease were found involving hematopoietic tissue, and these were both in Pacific white-sided dolphins. In the first case, there was splenomegaly, and marked enlargement of mesenteric nodes in an adult male animal. Histologic examination of various tissues from this animal revealed a somewhat unusual lymphoproliferative disease process. There was complete replacement of the normal architectural pattern of nodes and spleen by a somewhat polymorphic neoplastic lymphoid proliferation. There were areas of a predominance of lymphoblastic cells with round nuclei and prominent nucleoli. However, most areas consisted of a mixed population of small to medium sized lymphoid cells, and there was considerable deposition of collagenous connective tissue in the affected nodes with active fibroplasia (Figure 48). A scattering of eosinophils and plasma cells were present, and these were probably normal inhabitants or reactive cells. There were foci of lymphoid cells in the portal areas of some hepatic lobules. This was classified as a case of sclerosing malignant lymphoma.

The other hematopoietic neoplasm was quite unusual and worthy of description. An adult Pacific white-side dolphin beached and died. The necropsy examination was unremarkable except for moderately enlarged and turgid abdominal nodes, spleen and liver. Microscopic findings included a moderate reactive hyperplasia of the lymphoid germinal follicles in the spleen and nodes (Figure 49), with very active lymphoblastic activity in the center of the follicles. Of interest was the very marked proliferation and infiltration of eosinophilic granulocytes and precursors throughout the red pulp of the spleen, the cortex and medulla of the nodes and infiltrating out through the capsule of the nodes into the adjacent connective tissue (Figure 50), the portal triad areas of all lobules of the liver and in the outer cortices of both kidneys. In the red pulp of the spleen there was very active mitotic activity of the myeloblastic precursors (Figure 51) and evidence of maturation and differentiation through eosinophilic promyelocyte and myelocyte stages to mature eosinophils (Figure 52). Immature eosinophil precursors were also present in the infiltrates of the nodes and liver. Realizing that these animals normally have an elevated peripheral eosinophil count and moderate eosinophil presence in lymphoid tissues, nevertheless this was believed to be a neoplastic proliferation of the eosinophil granulocyte, and leukemia-eosinophilic or perhaps something akin to the generalized eosinophil proliferative disorders that have been described in man.[42,43] Unfortunately, bone marrow or peripheral blood were not available for evaluation from this animal.

VIII. MISCELLANEOUS

An adult California sea lion was found beached and died without opportunity for clinical evaluation or supportive therapy. At necropsy examination, a mass was found in the abdominal cavity, along the mid-line adjacent to the left kidney. There was an infiltrative mass in that kidney as well. There was some patchy pneumonia present, and other organs were normal.

Microscopically, the abdominal mass and renal lesion were the same and were found to be a neoplasm which was composed of small to medium sized tumor cells with hyperchromatic nuclei and clear cytoplasm, arranged on a fine reticular stroma in a prominent organoid pattern typical of an apudoma or neuroendocrine carcinoma (Figure 53). No tumor lesions were found other than at the sites noticed grossly. A Grimelius stain demonstrated relatively few silver staining cytoplasmic granules in some cells, while most cells were devoid of granules. This tumor was considered to be a neuroendocrine carcinoma (apudoma), arising adjacent to and metastasizing to the kidney.

We found a well differentiated fibroma of the gingiva in a white-sided dolphin.

IX. DISCUSSION

This report has added a considerable number of previously unreported neoplasms in marine mammals. Considering a base of approximately 1500 animals, the 39 neoplasms detected would mean that the total incidence of neoplastic disease in this group of animals was over 2.5%. Twenty-eight were malignant and probably led to the death of the animal, or a cancer related death rate of just over 1.9%. This is quite low in comparison to man in which the cancer related death is given as 17.8%.[44] We must consider the likelihood, however, that neoplasms may be more common in marine mammals than this study would indicate, but that many of those animals that die of malignancies are never recognized. It is a disease of older animals, and these tend to stay in more secluded areas away from human population centers, and are less apt to become beached at a location which is frequented by humans. Others may simply drown or be destroyed by predators at sea.

It is of considerable interest that our study yielded such a high percentage of malignancies of tissues such as the lung, biliary system, and genitourinary system in those California sea lions beached along southern California beaches. Carcinomas of the lung and biliary system have not previously been described in pinnipeds. The occurrence of epidermoid carcinomas of the lung is quite unusual since epidermoid carcinoma of the lung is one of the least frequent types of primary lung cancer in animals,[45] in contrast to man where it is so common and associated with cigarette smoking.

Both experimentally and epidemiologically, it has been shown that tissues such as lung, liver, and urinary tract may act as target organs for the induction of cancers caused by carcinogenic chemicals. We do know that there has been considerable pollution of the offshore waters along southern California coasts, and one must wonder if this could be a factor in the development of some tumors in these animals.

The presence of considerable amounts of organochlorine type pollutants has been repeatedly documented in these coastal waters,[46,47,48] and it has been shown that in such an environment there is increasing concentrations of some pollutants, particularly organic mercury and high molecular weight chlorinated hydrocarbons, in increasing trophic levels of the food web.[48] This would make it feasible for these marine mammals to be chronically exposed to low levels of chemicals in their diet. Organochlorine pollutant tissue levels have been evaluated in southern California marine mammals, and one study concluded that these levels were associated with premature births in sea lions.[49] Fish from these same waters have been shown to have fin erosion disease and other anomalous disorders, caused by the discharge of chemical pollutants through sewage treatment plants.[50]

In addition to these well documented pollutant levels in southern California waters, we don't know the possible impact of pollutants associated with oil spills such as occurred at Santa Barbara in 1969. We do know that there are significant levels of known carcinogen introduced into the marine environment with such events. Following the explosion of the tanker Sansinena in Los Angeles Harbor in 1977, some studies were conducted of the mollusks in the area, and tissue levels of benzopyrene in those animals were found to be markedly elevated.[51] Tissue burdens of benzopyrene have been associated with neoplastic proliferative disorders in mussels along the Oregon coast,[52] but similar studies have not been done along the Los Angeles area. It has been suggested that increased incidences of fish tumors might function as an "early warning system" for the presence of carcinogens in aquatic environments,[53] and in this regard skin papillomas were found to be increased in southern California waters, but to a very minimal degree as compared to the Washington coast where the incidence is high. Whether the papillomas were caused by a specific pollutant has not been elucidated.

Generalized adenocarcinoma was recognized in two California sea lions that beached and died in close proximity to one another,[26] and it was speculated that they may have had a similar or common exposure to an oncogenic factor.

It seems reasonable to speculate that there are cases of carcinogen induced neoplasms in our group of marine mammals, particularly because of the unusual occurrence of carcinomas of the lung, biliary tract, and urogenital tract. In particular, the obvious relative high percentage of tumors of the genital tract bears further investigation, in the face of the known environmental contamination by organochlorine compounds, and what we now understand of some of the effects on the steroid hormones. DDT analogs, present in DDT mixtures have been shown to have estrogenic activity, and inhibit in vitro binding of estradiol to uterine receptors.[56]

PCB, another common pollutant in the environment, was found to have estrogenic effect and augment progesterone synthesis in rats,[57] and in another study was found to have gonadal effects in cod fish.[58] Benzo(a)pyrene, a petroleum related pollutant in our area, has a marked affinity for binding to the DNA of human endometrial tissue in culture.[59]

The possible long-term impact of these compounds on the marine mammal along polluted coastal areas is obvious, as these animals are exposed to chronic low level ingestion of these pollutants in their diet. Whether these compounds with hormonal effects play any role in the long-term carcinogenesis of the genital system of these animals remains to be seen, but it is certain that further research on this problem must be carried out.

To prove a cause/effect relationship, one must have a large population exposed to a common known carcinogen, and study this population for a long period, comparing the resultant data with a comparable control population not exposed to the carcinogen. It is doubtful that such a situation can exist, particularly at a time when research funds for this type of study are meager. An alternative method of study would be to have a large collection of captive animals, chronically exposed to whatever carcinogens they would normally be exposed to in their natural habitat, and observe them throughout a long period of time. This type of study is not feasible either.

We have made certain preliminary attempts to quantitate tissue residue levels of certain organochlorine pollutants that we know are prevalent in some areas of the southern California coast, and this data is presented in another chapter under toxicologic studies. Unfortunately, there are many variables such as migration patterns of these animals, precisely where had it been living and what levels of dietary pollution was it exposed to, metabolic and excretory pathways for these animals, residue levels in relation to age, sex, nutritional status (body fat), pregnancy, nursing behavior, and many other factors. This makes the evaluation of resultant data extremely difficult, or perhaps impossible to interpret and try to relate to a condition such as neoplasia.

Nevertheless, we do feel that statistically there is evidence that we have been seeing an increased incidence of certain types of neoplasia which is probably caused by environmental chemical pollution in this area.

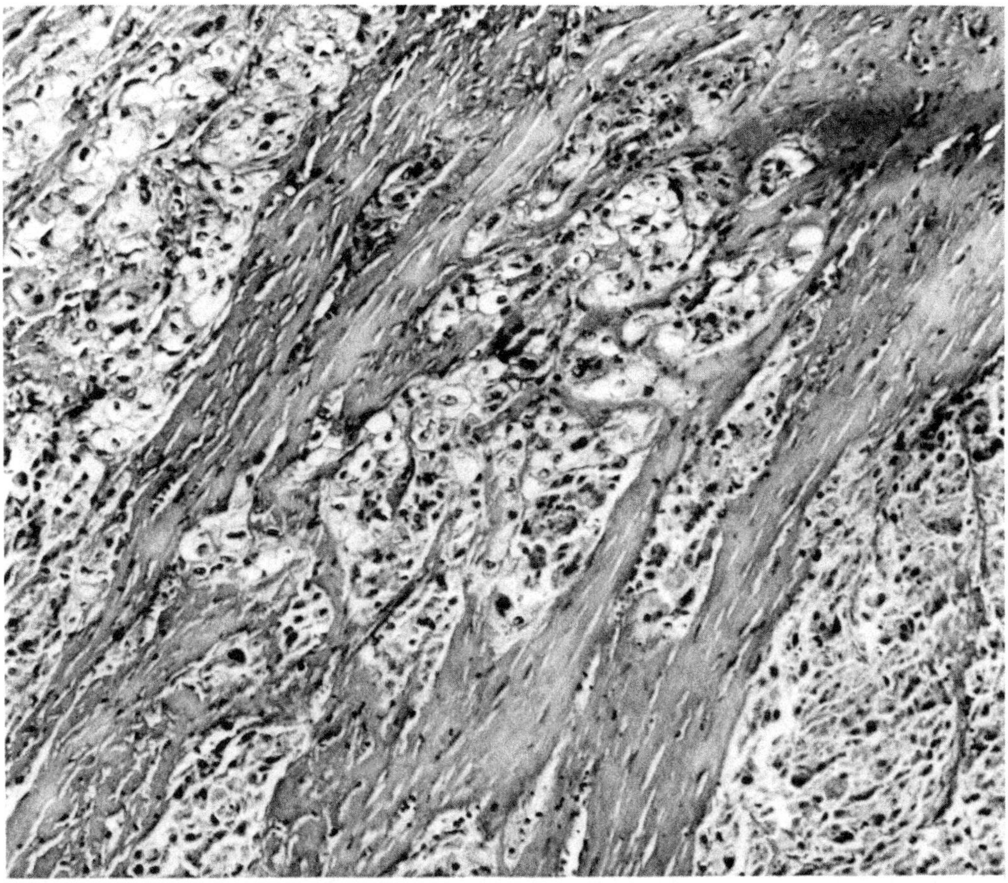

FIGURE 1. Section from a tumor in the subcutis of the neck of a California sea lion, probably arising from sweat glands of the skin. The tumor cells have a mucoepidermoid carcinoma appearance. (From *Mammals of the Sea*, Charles C Thomas Publisher, 1972.)

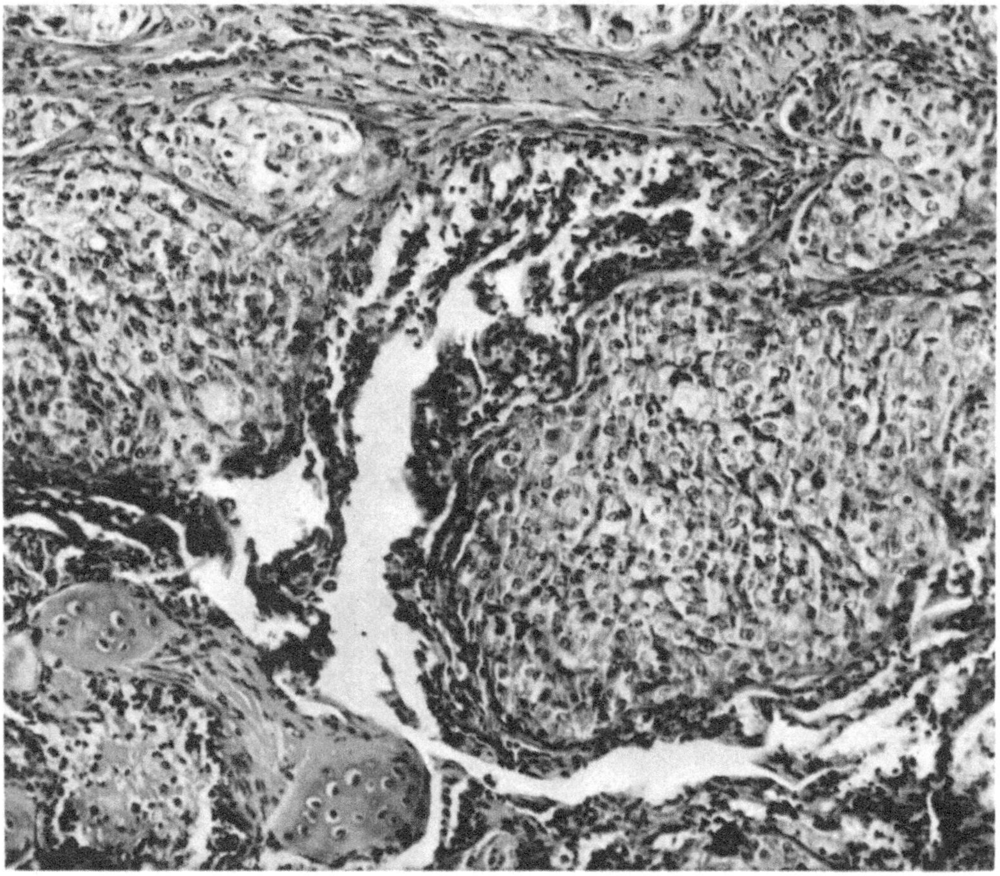

FIGURE 2. Lung section from the same sea lion as Figure 1, showing a metastasis from the skin primary. (From *Mammals of the Sea*, Charles C Thomas Publisher, 1972.)

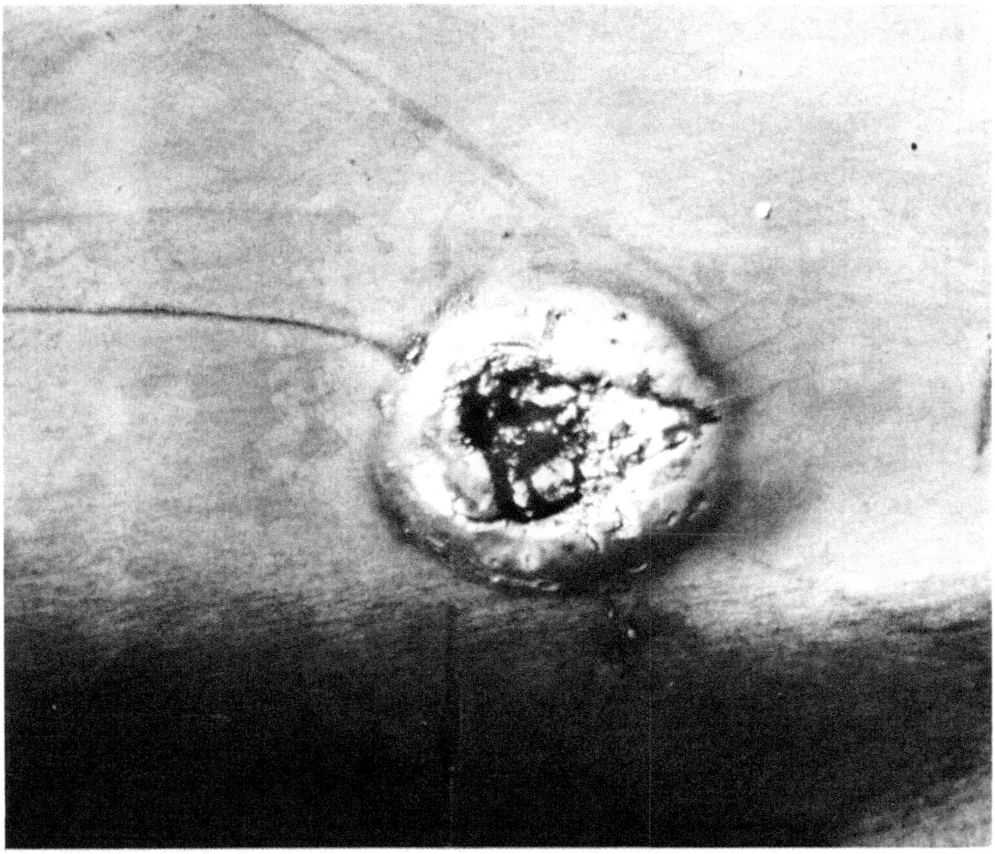

FIGURE 3. Gross picture of a dolphin, showing an ulcerated and necrotic lesion on the side, which was found to be a squamous cell carcinoma on biopsy. The lesion was cured by cryosurgery.

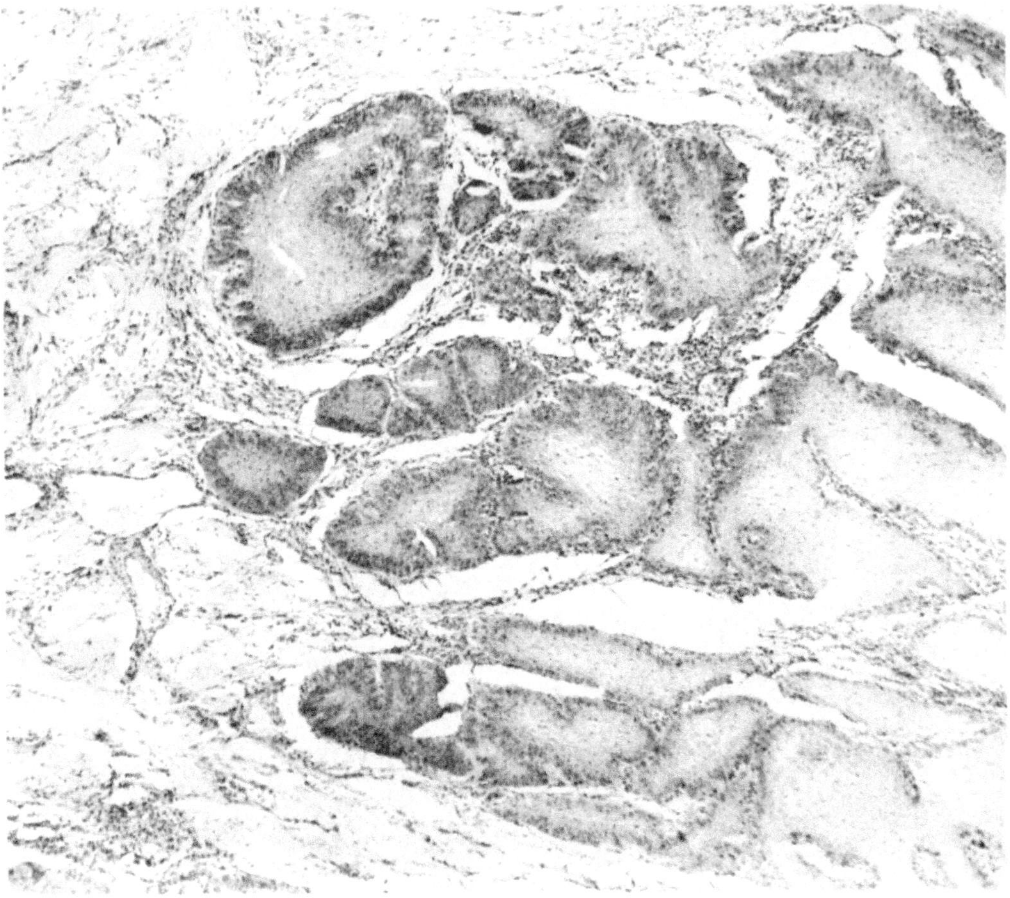

FIGURE 4. Biopsy detection from the lesion shown in Figure 3, demonstrating the deeply infiltrating cords of well differentiated squamous cell carcinoma into the dermis.

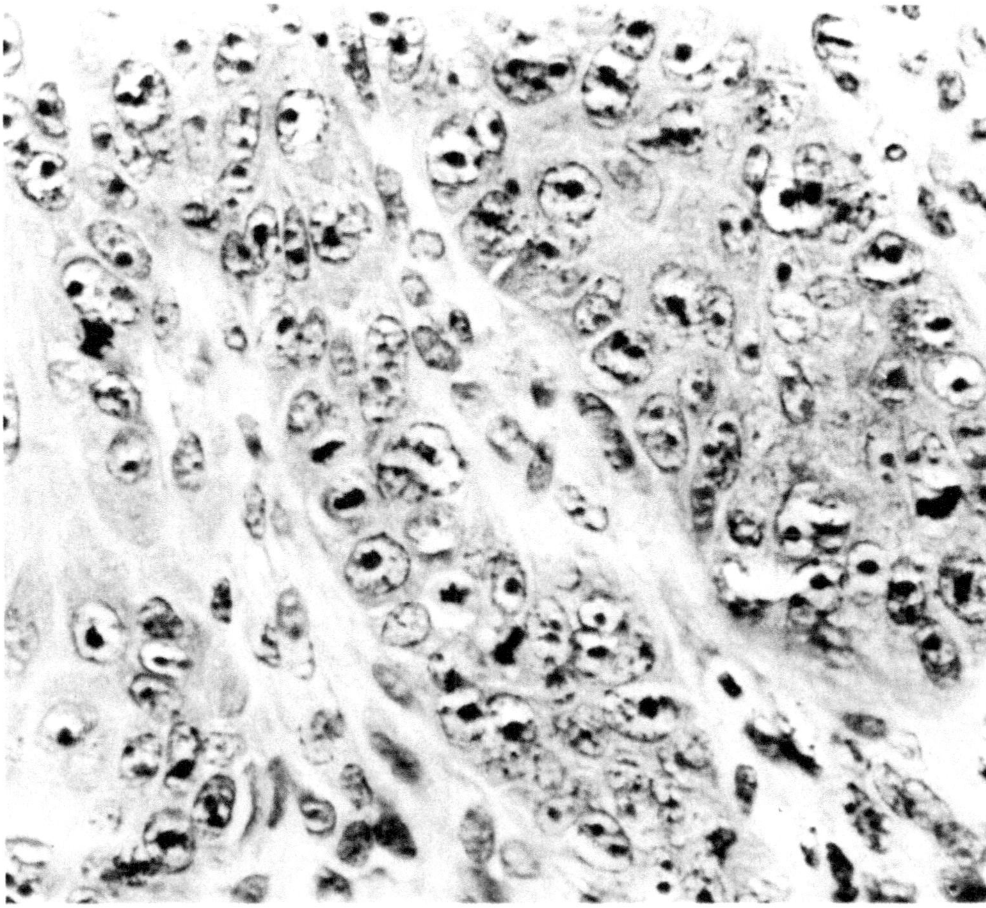

FIGURE 5. Higher magnification of the squamous cell carcinoma from the side of a dolphin, showing marked cellular atypia and several mitotic cells in the field.

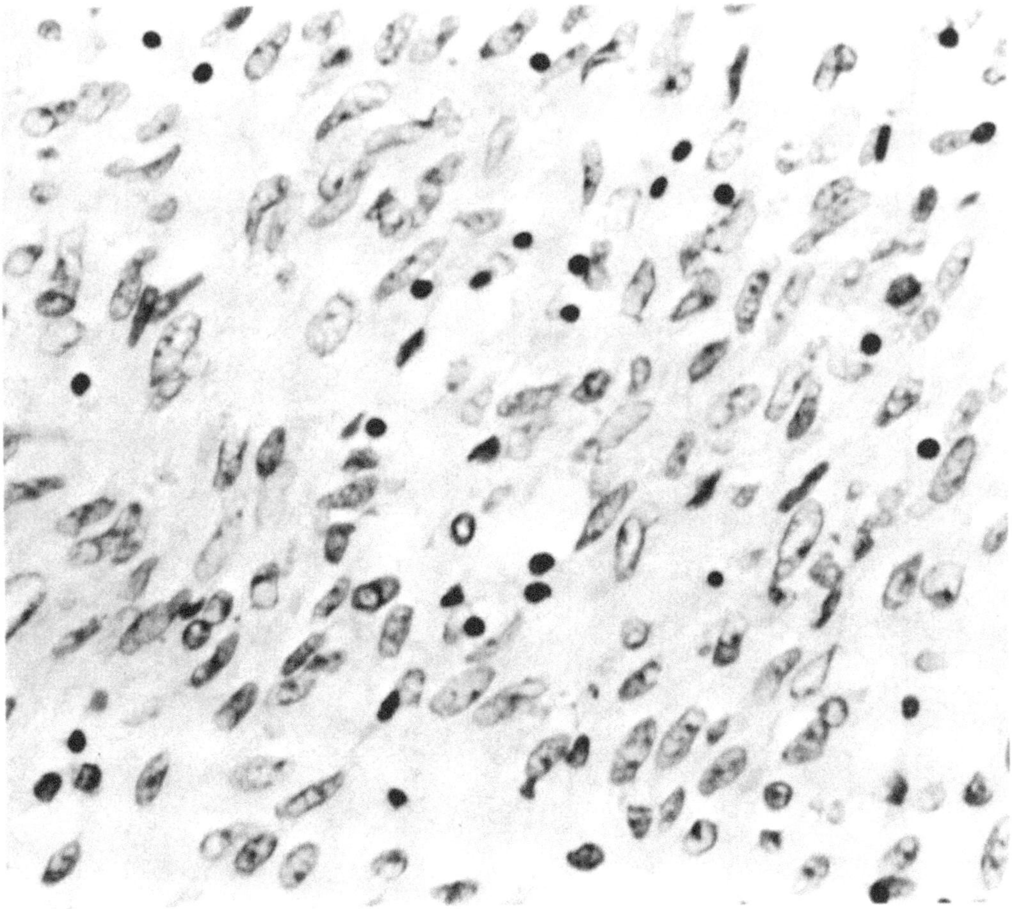

FIGURE 6. Section of a firm dermal nodule from a mature female California sea lion. This is a well differentiated leiomyoma. Note the sausage shaped nuclei with nuclear chromatin pattern typical of smooth muscle cells.

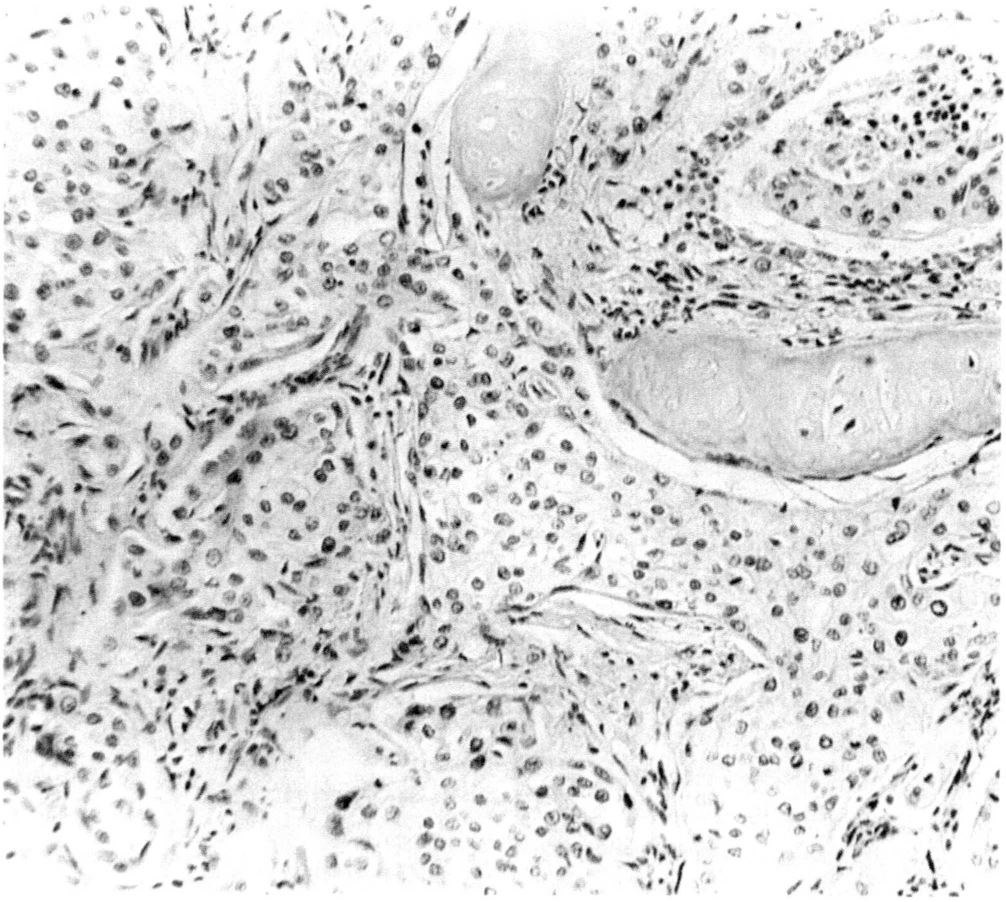

FIGURE 7. Lung section from an adult female sea lion which had a moderately differentiated epidermoid carcinoma of the lung. There are sheets of squamous epithelial cells filling alveoli and bronchiole, but there are no keratinized areas or epithelial pearls seen.

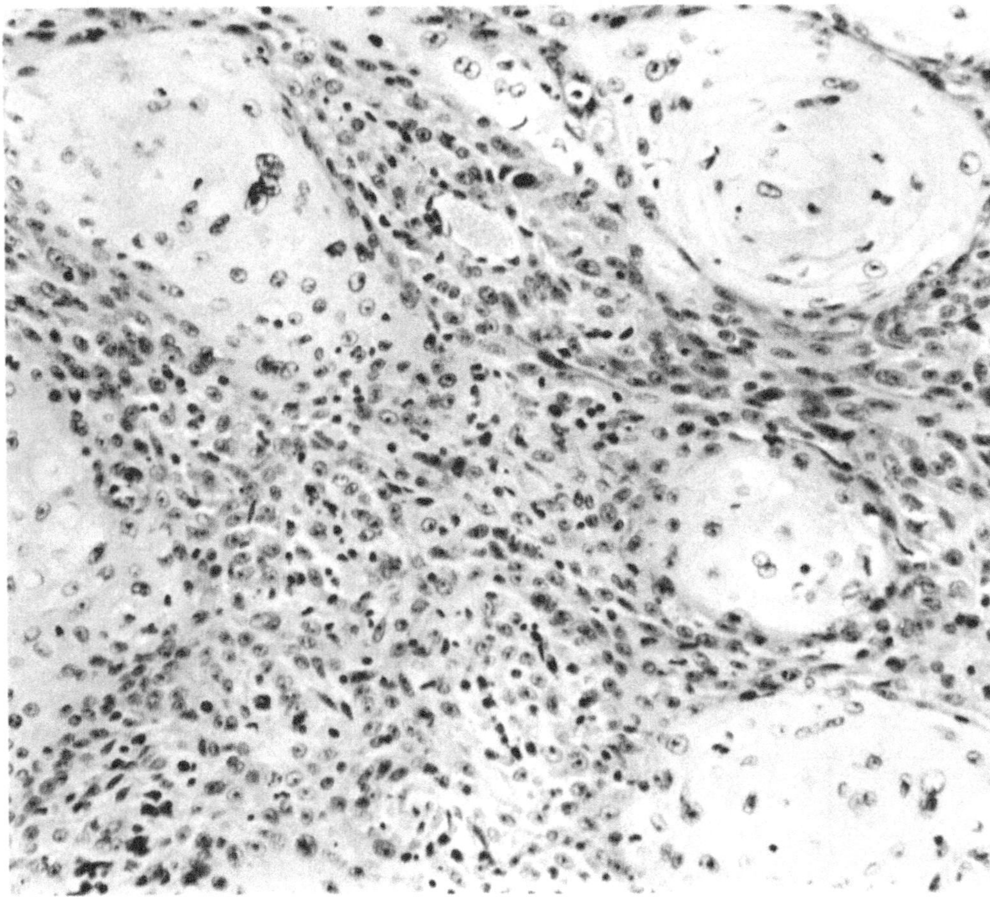

FIGURE 8. This is a section from a well differentiated epidermoid carcinoma of the lung of a mature female sea lion. Note the keratin pearls and sheets of less differentiated cells between them.

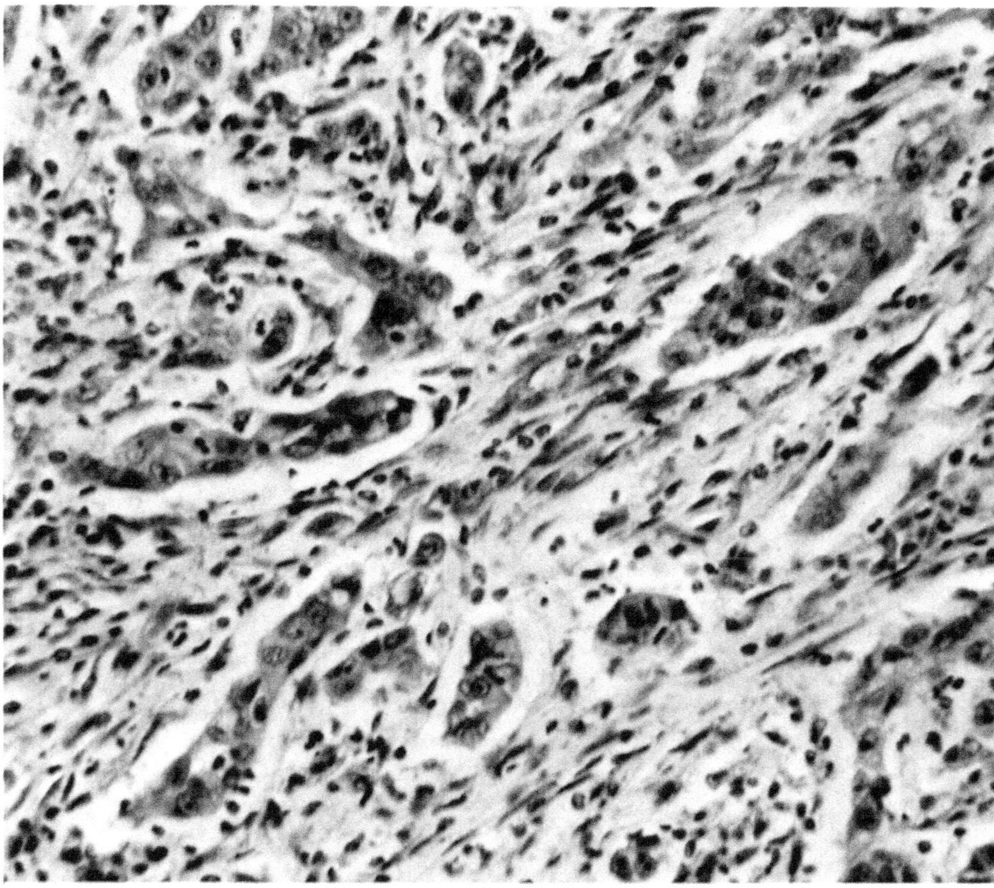

FIGURE 9. Section from a poorly differentiated adenocarcinoma of the lung of a sea lion. There is a suggestion of acinar formation by the tumor cells, with considerable fibroplasia and some inflammation.

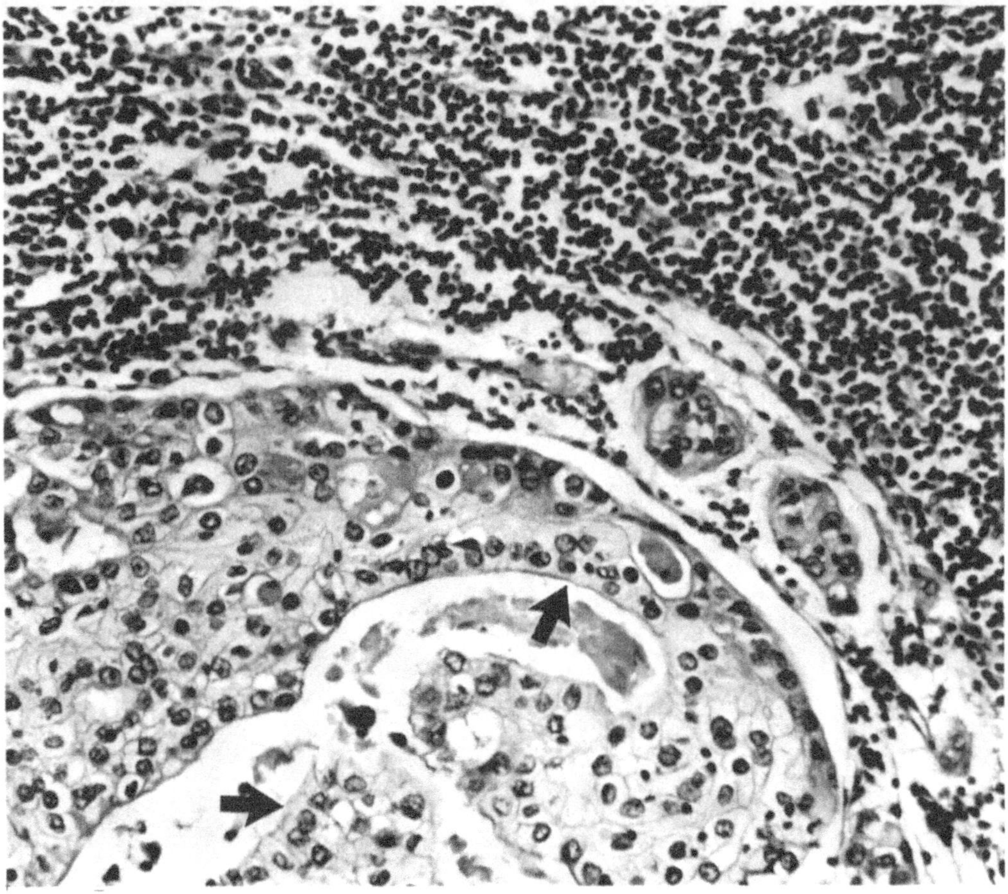

FIGURE 10. Bronchial lymph node from the sea lion in Figure 9, demonstrating metastasis of the adenocarcinoma cells from the lung. There is squamous metaplasia of the cells, but cuboidal to low columnar tumor cells can be seen lining an alveolar-like space (arrows).

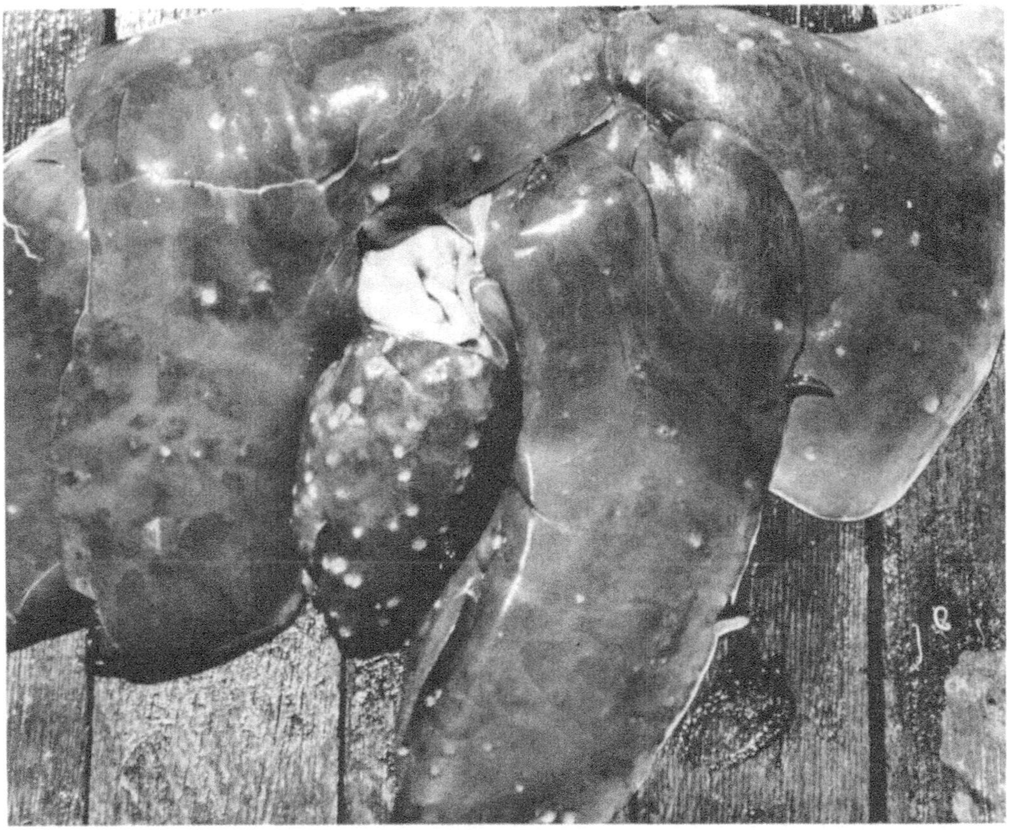

FIGURE 11. Liver from a 5-year-old male sea lion, showing numerous white focal lesions. This was found on histologic examination to be a bile duct adenocarcinoma.

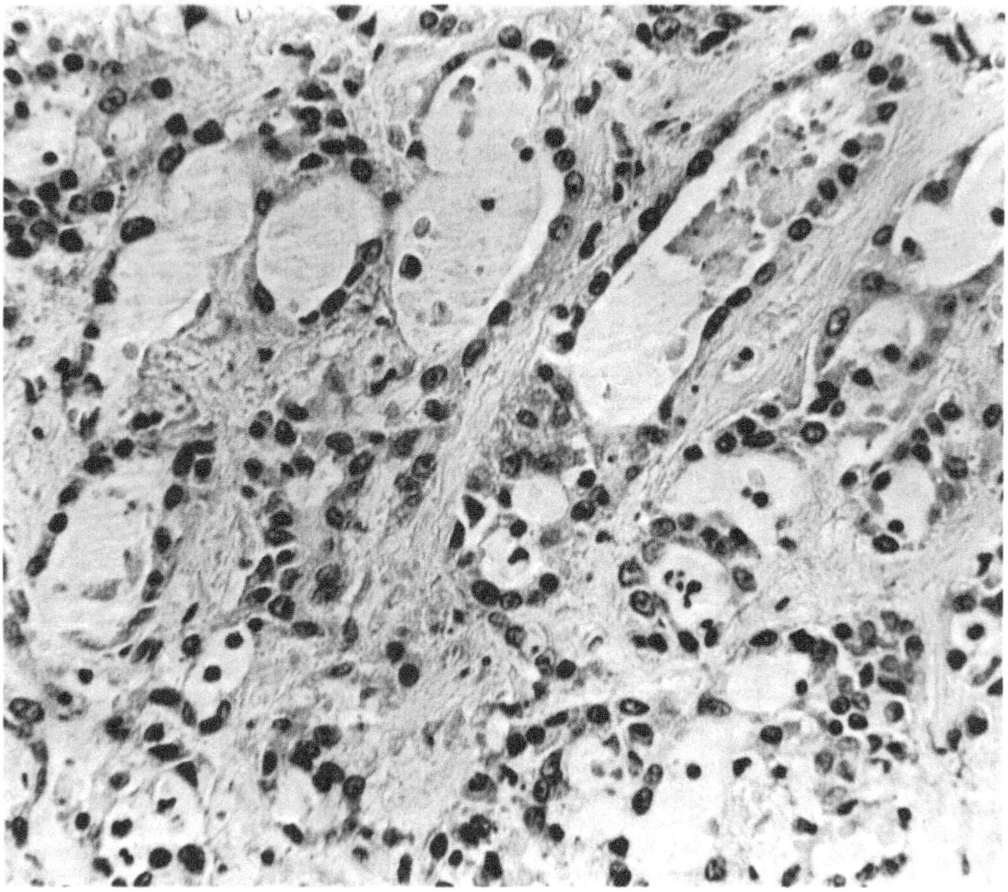

FIGURE 12. Histologic section of the bile duct adenocarcinoma shown in Figure 11. Note the well differentiated low cuboidal tumor cells lining mucin filled duct like spaces.

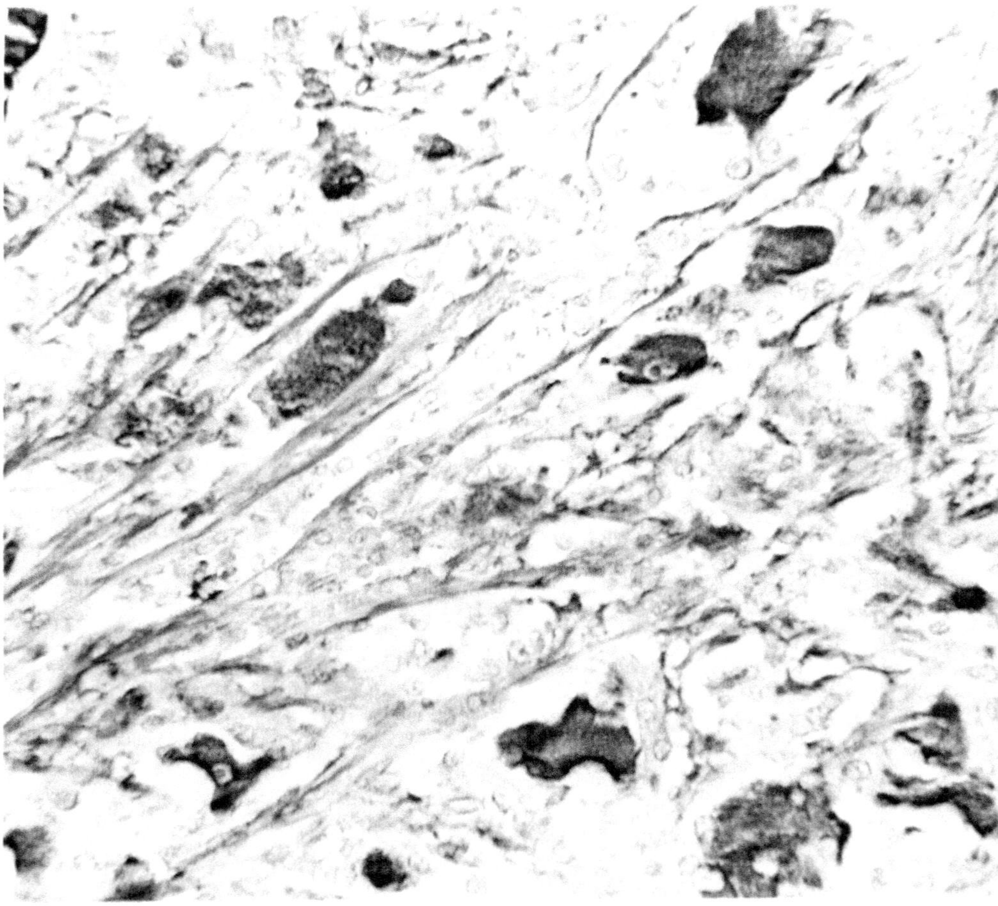

FIGURE 13. This is a PAS stained section of the bile duct adenocarcinoma shown in Figure 12, demonstrating a moderate amount of PAS reactive mucin material.

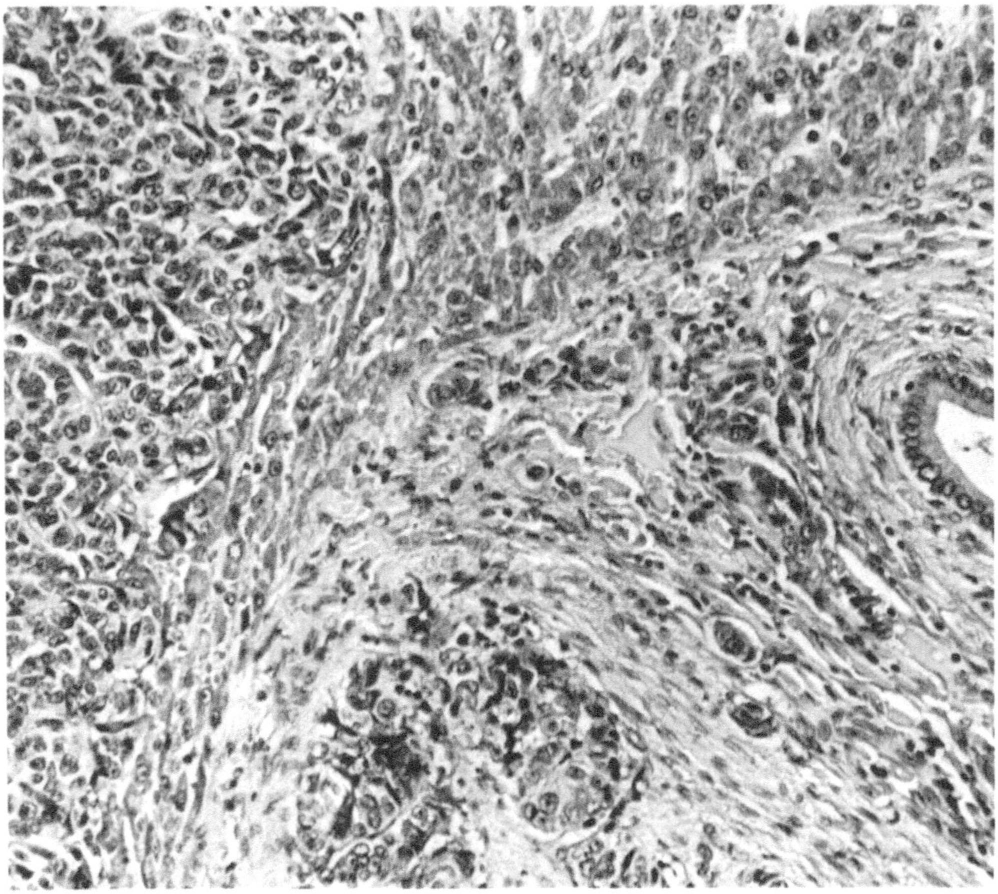

FIGURE 14. Sheets and nests of tumor cells, encroaching on liver parenchyma toward a bile duct, are seen in this section of an undifferentiated bile duct adenocarcinoma found in an adult female sea lion.

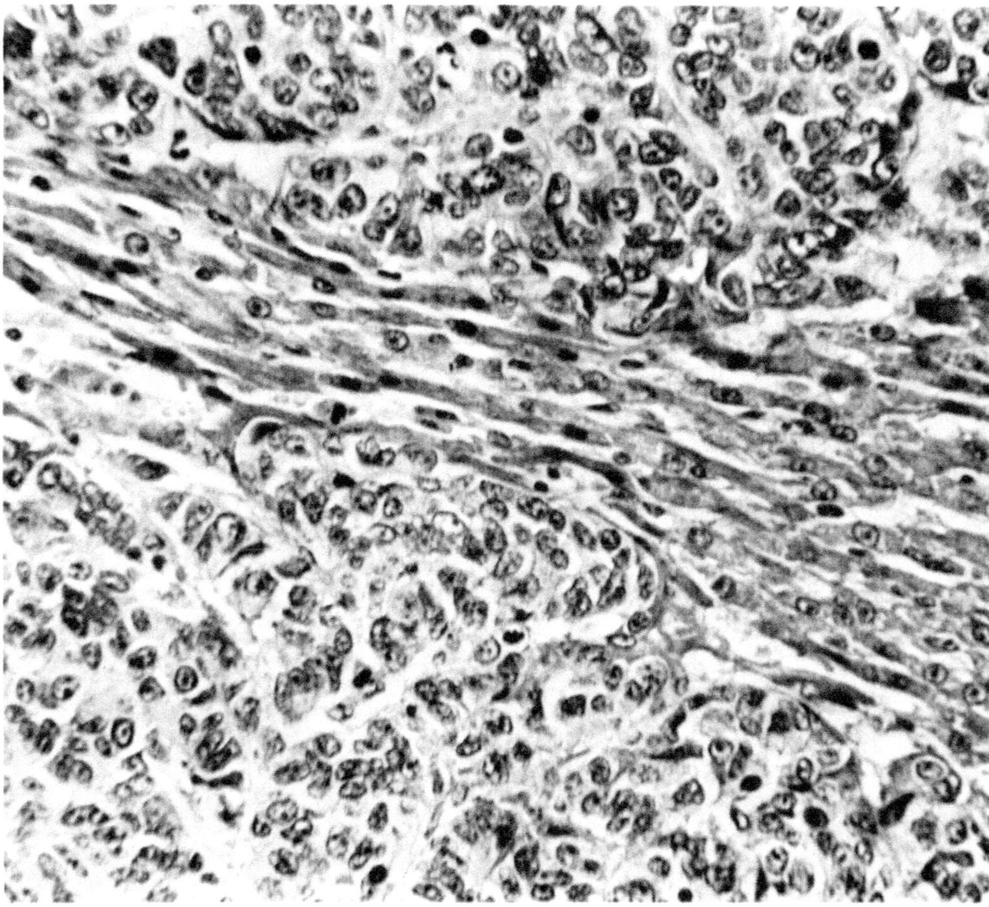

FIGURE 15. Solid sheets of bile duct adenocarcinoma tumor cells invading along either side of a zone of compressed liver cord cells seen traversing through the center of the field. This tumor was from an adult female sea lion.

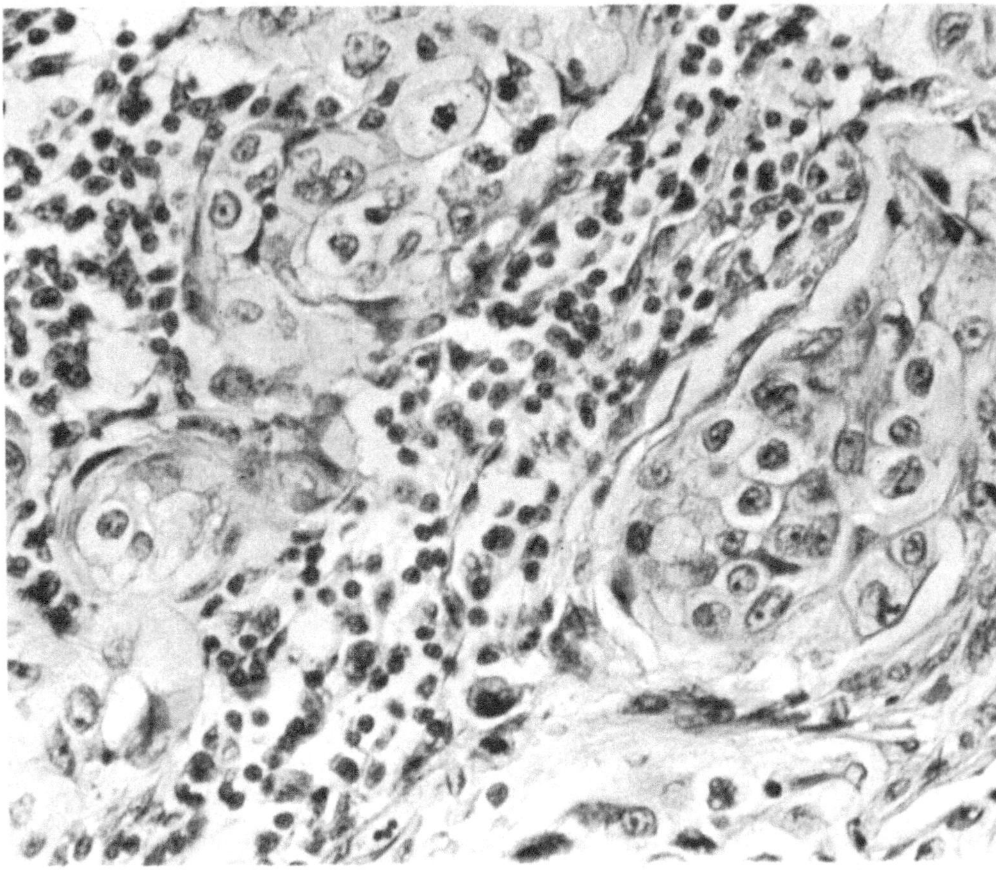

FIGURE 16. Nodal metastasis from an anaplastic bile duct adenocarcinoma, showing marked squamous metaplasia of the tumor cells.

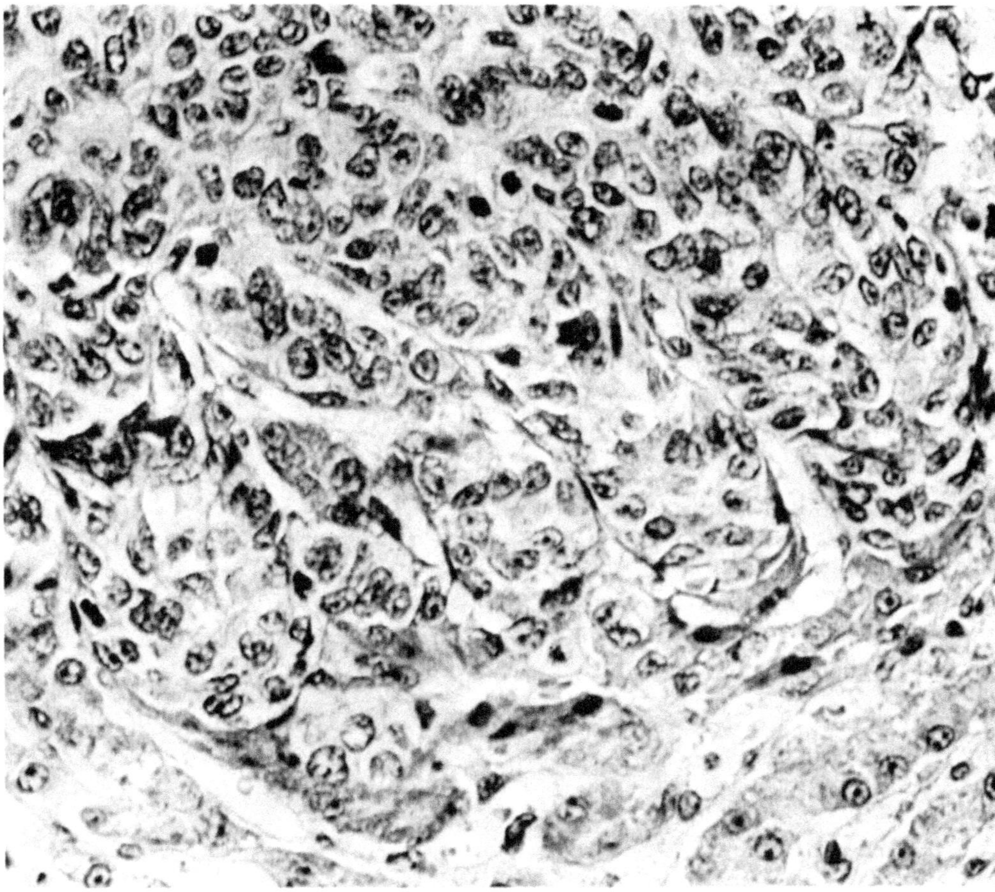

FIGURE 17. Bile duct adenocarcinoma compressing liver cords (lower right and left). The tumor cells are large and pleomorphic nuclear pattern is prominent. The tumor cells have abundant pale to clear cytoplasm.

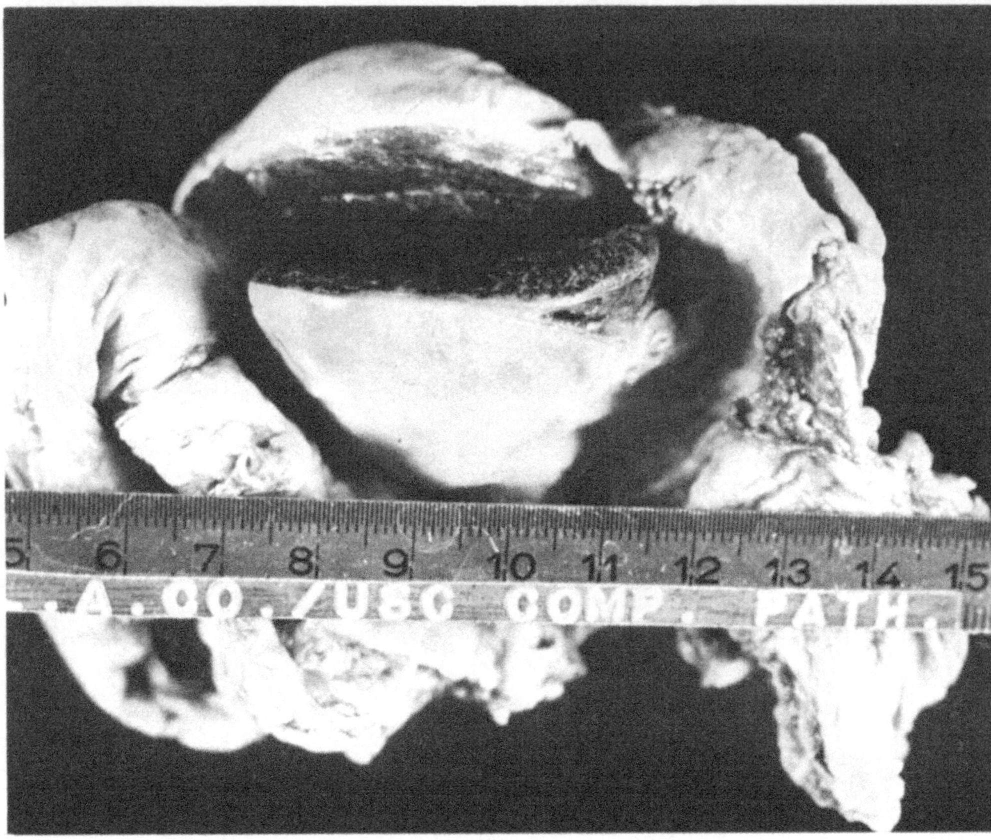

FIGURE 18A. A large firm mass involving the body of the uterus of an adult sea lion was found to be a benign leiomyoma.

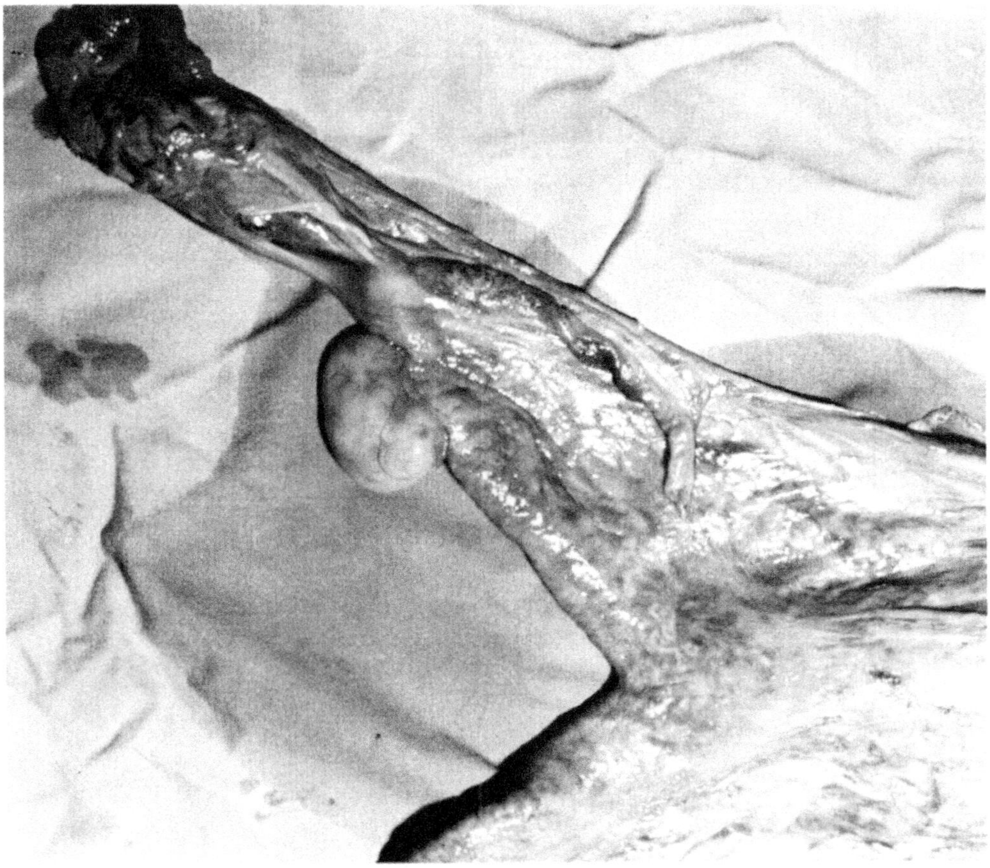

FIGURE 18B. Uterus from a sea lion showing a small tumor mass of the right uterine horn, which was a leiomyoma.

FIGURE 19. Tissue section from a benign uterine leiomyoma, showing parallel bundles of smooth muscle cells with plump sausage shaped nuclei.

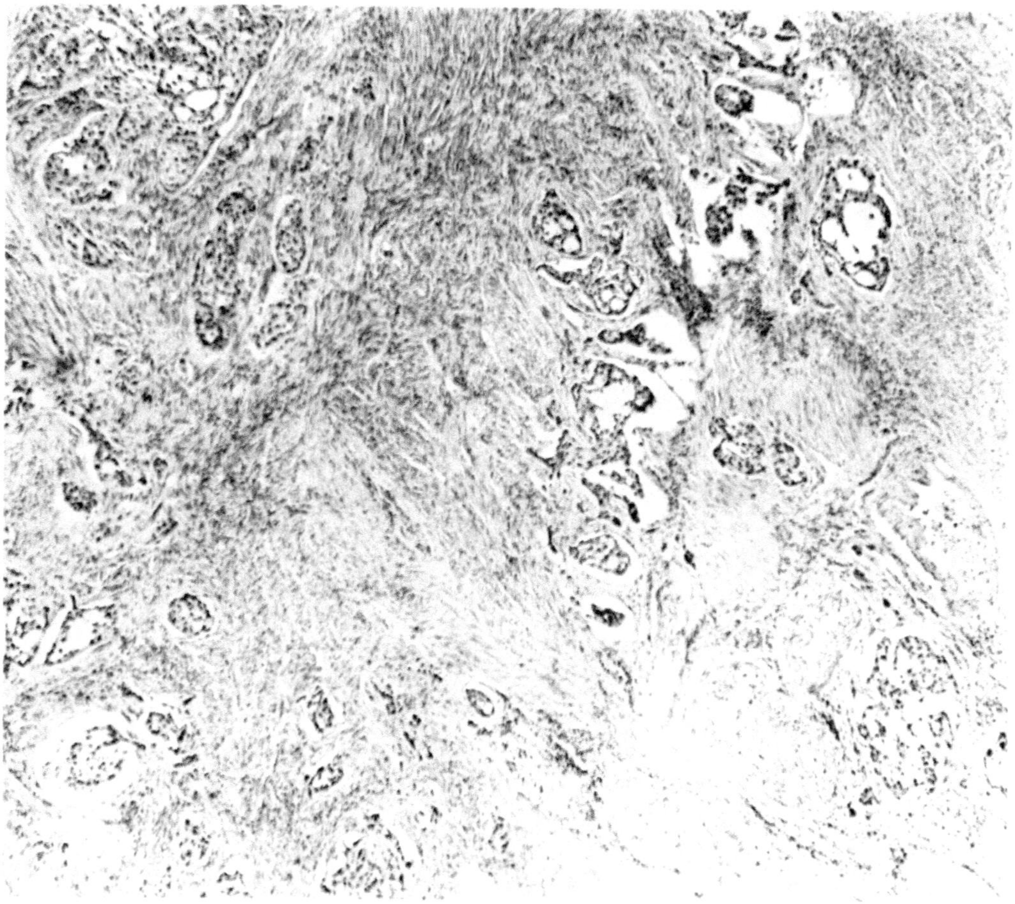

FIGURE 20. Section of uterine wall from a sea lion with a uterine adenocarcinoma. Note the nests of tumor cells extensively invading the muscle wall, and some are forming a glandular pattern.

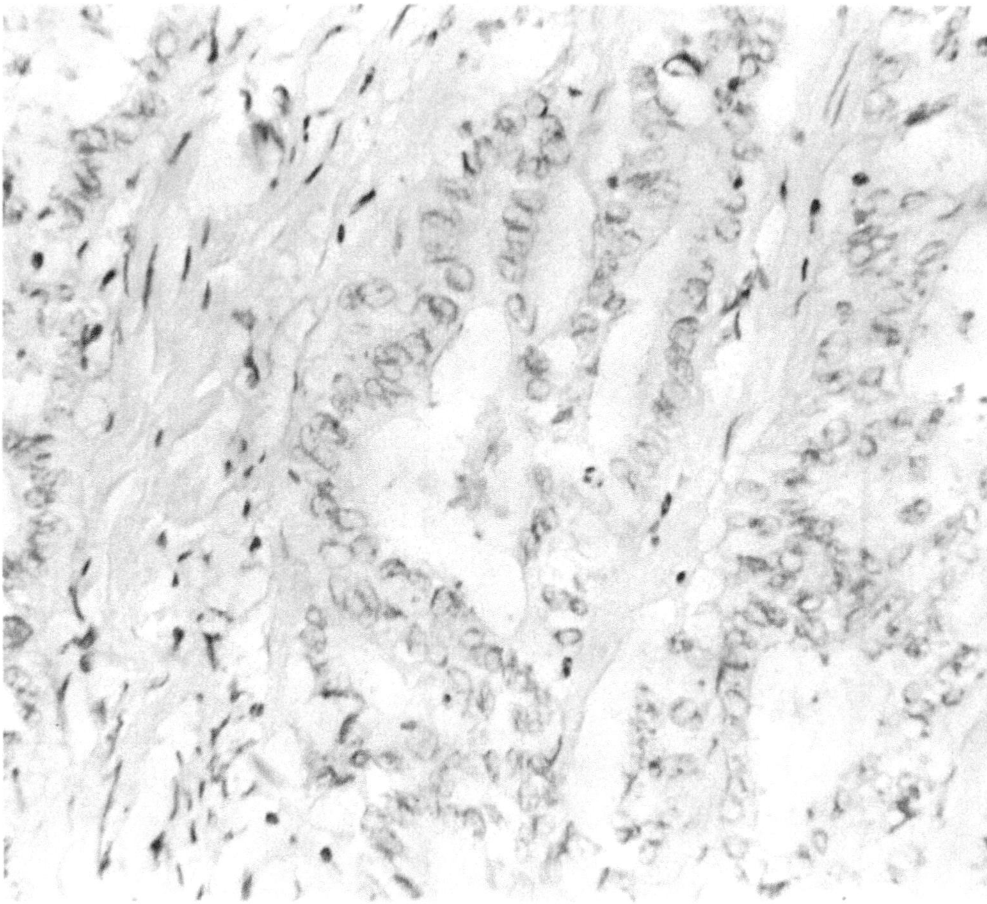

FIGURE 21. Higher magnification of the section shown in Figure 20, showing well differentiated uterine adenocarcinoma invading the muscle wall, and a glandular pattern is apparent.

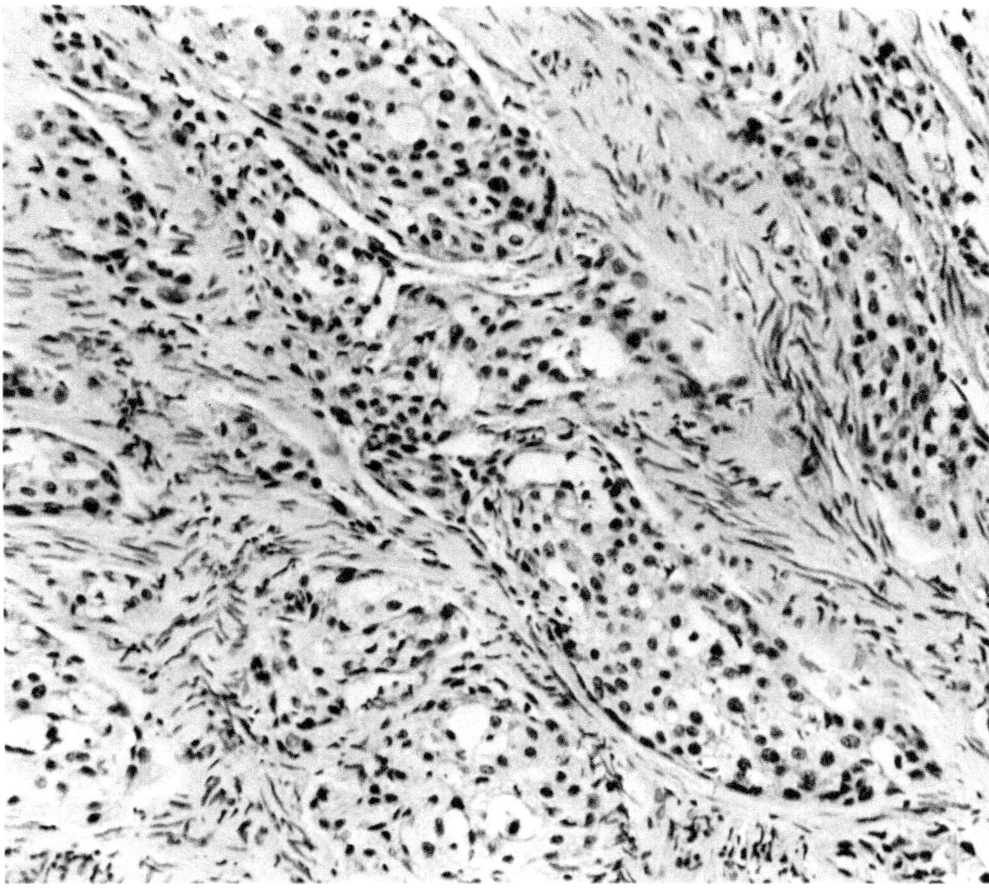

FIGURE 22. Section from an undifferentiated uterine adenocarcinoma found in an old sea lion. Nests of tumor cells can be seen invading between muscle bundles.

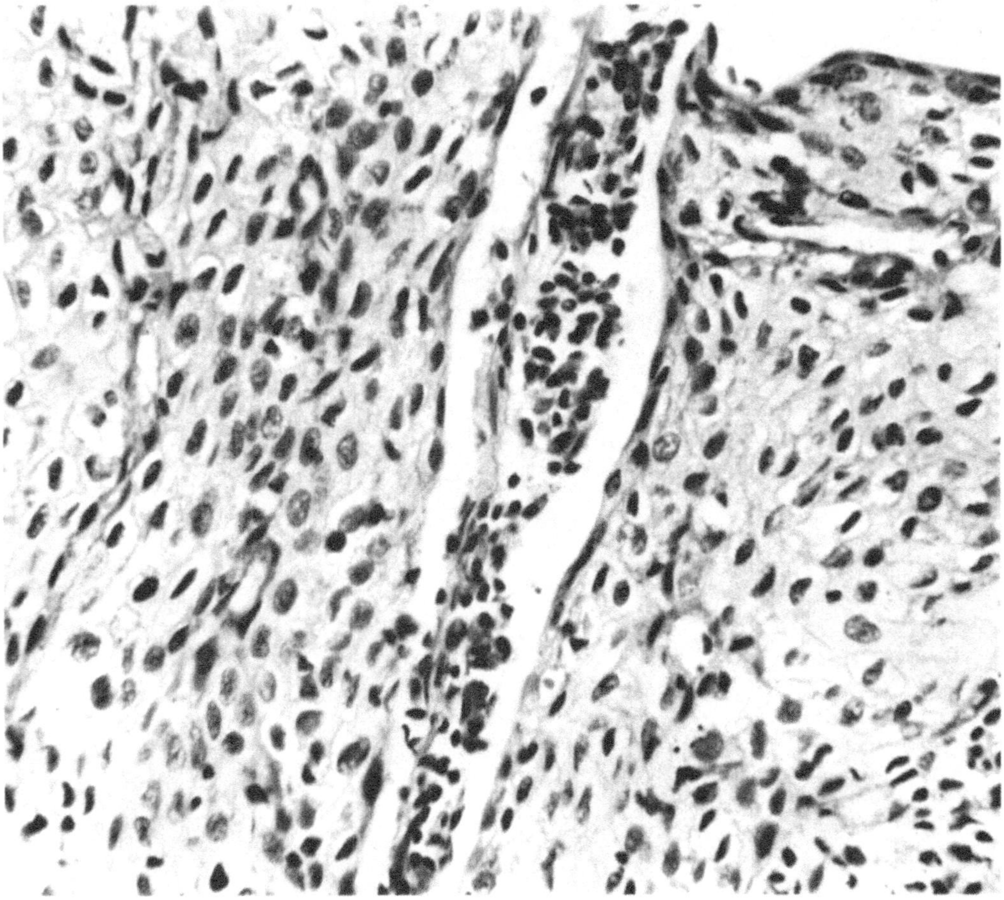

FIGURE 23. Higher magnification of tumor cell nests from Figure 22, showing the moderate degree of squamous metaplasia of the tumor cells.

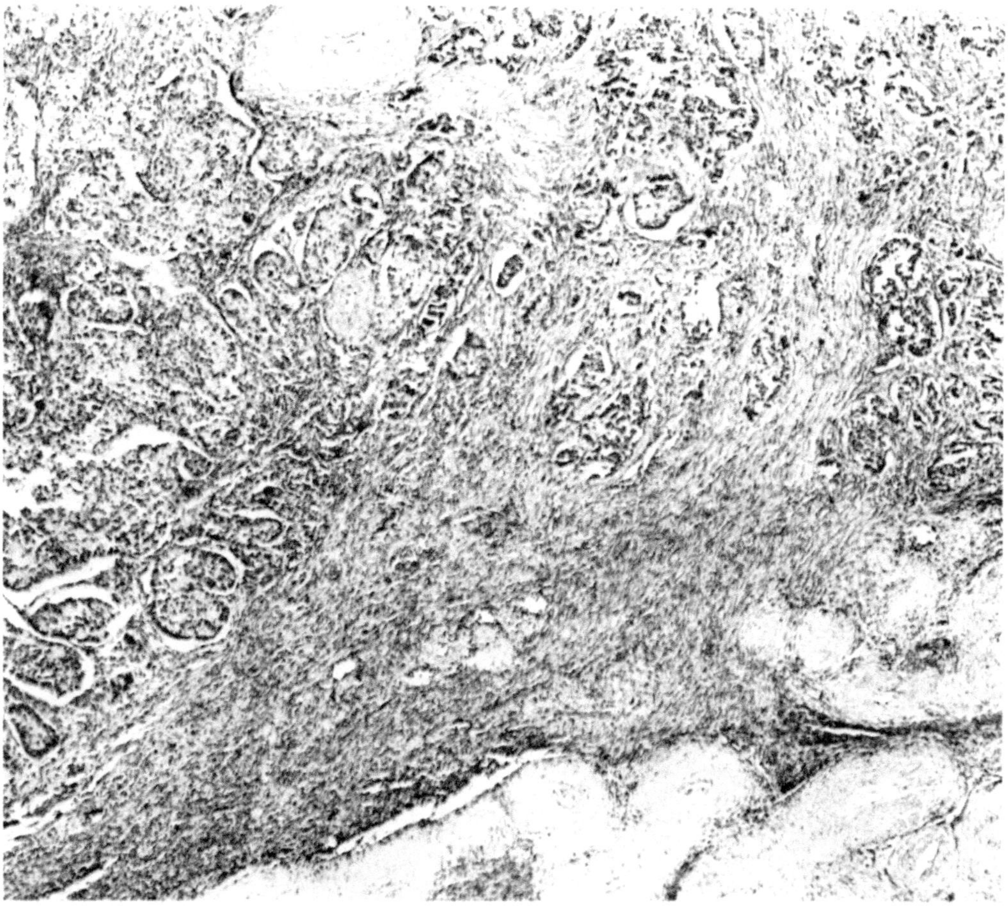

FIGURE 24. Section of ovary from an adult sea lion, showing extensive stromal invasion by a moderately differentiated adenocarcinoma.

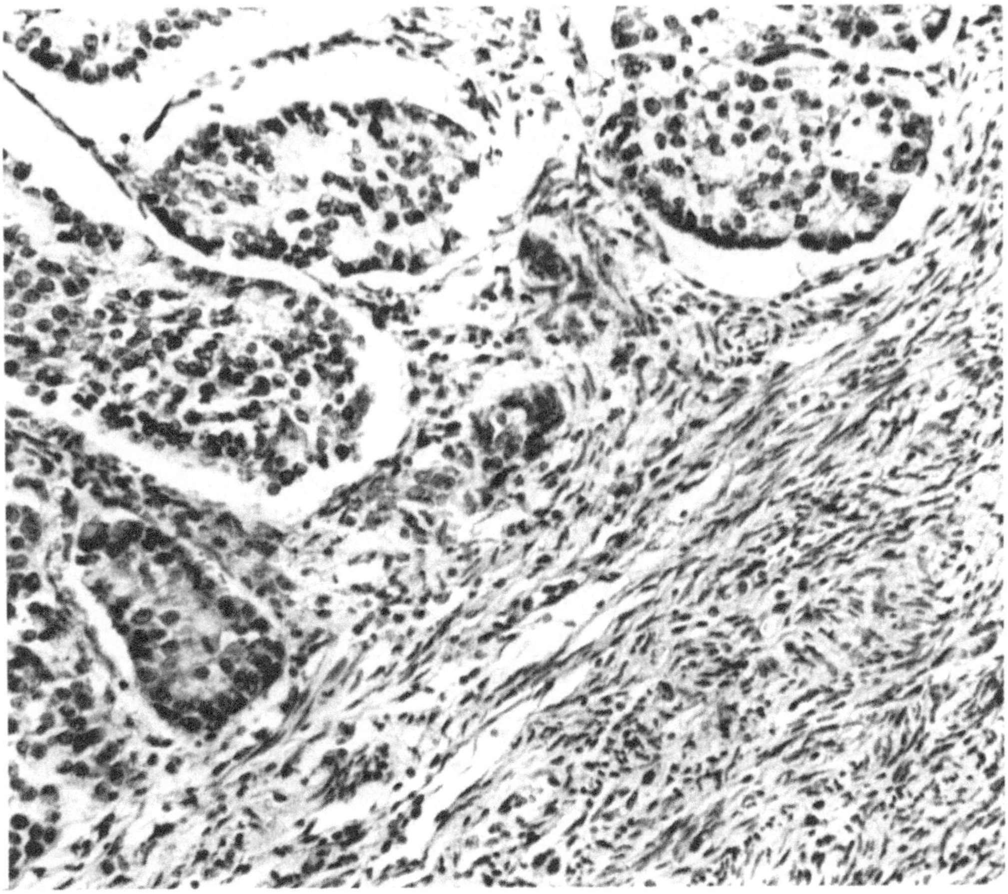

FIGURE 25. Higher magnification of the ovarian adenocarcinoma shown in Figure 24, demonstrating a pseudoglandular pattern of the tumor cells that are invading deep into the ovarian stroma.

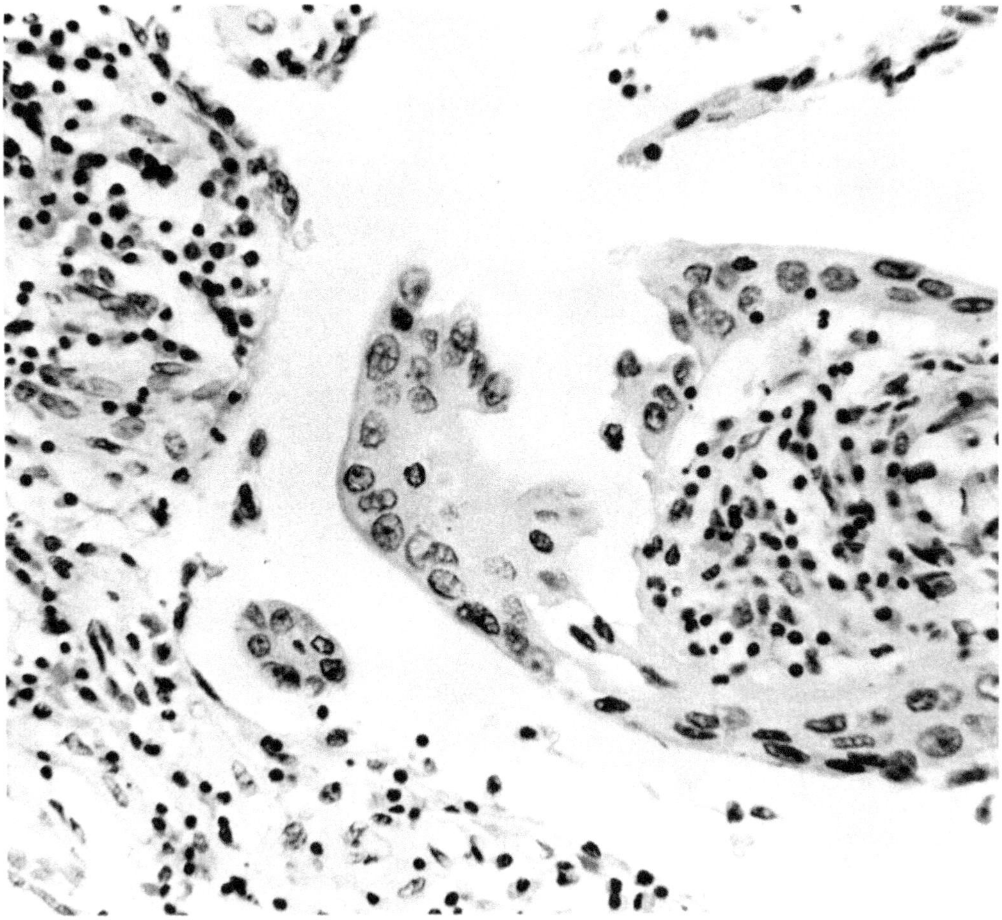

FIGURE 26. Peritoneal surface from a sea lion with an ovarian adenocarcinoma, in which there was extensive abdominal seeding by the tumor cells.

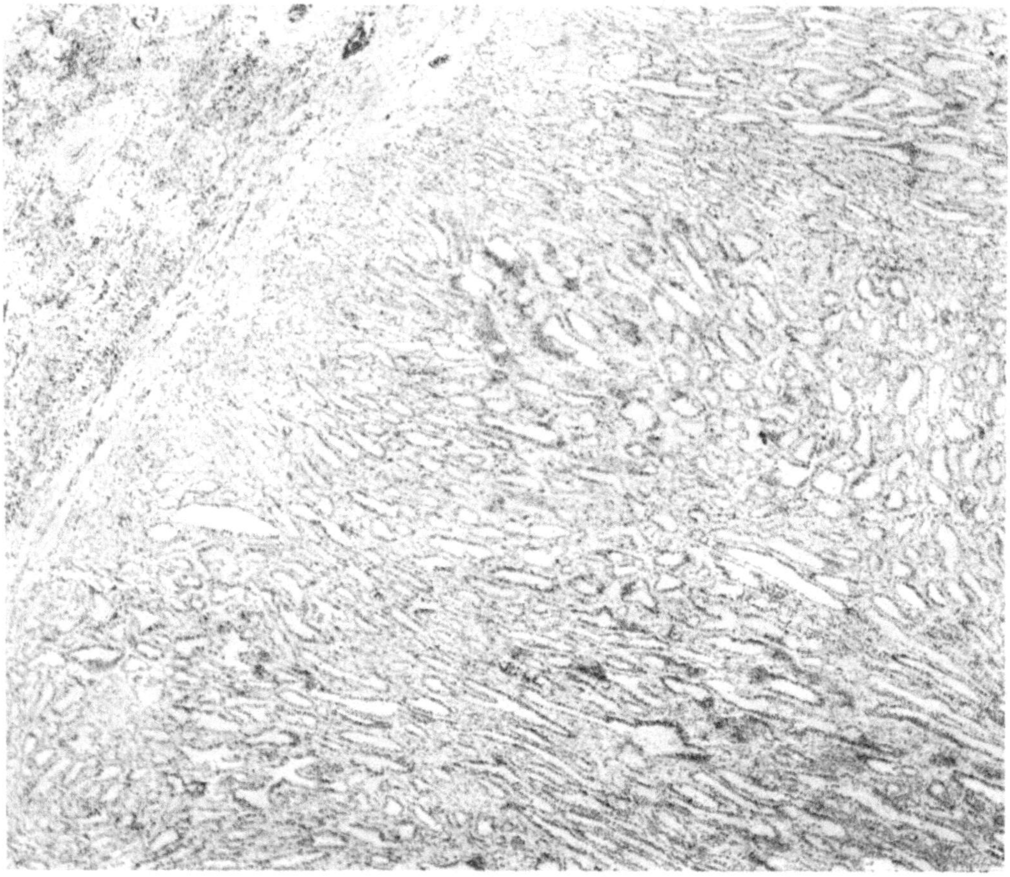

FIGURE 27. Section from an ovarian adenoma, showing the well differentiated follicular pattern extending out to the capsule. The normal ovarian tissue had been compressed into a narrow rim around one side of the tumor.

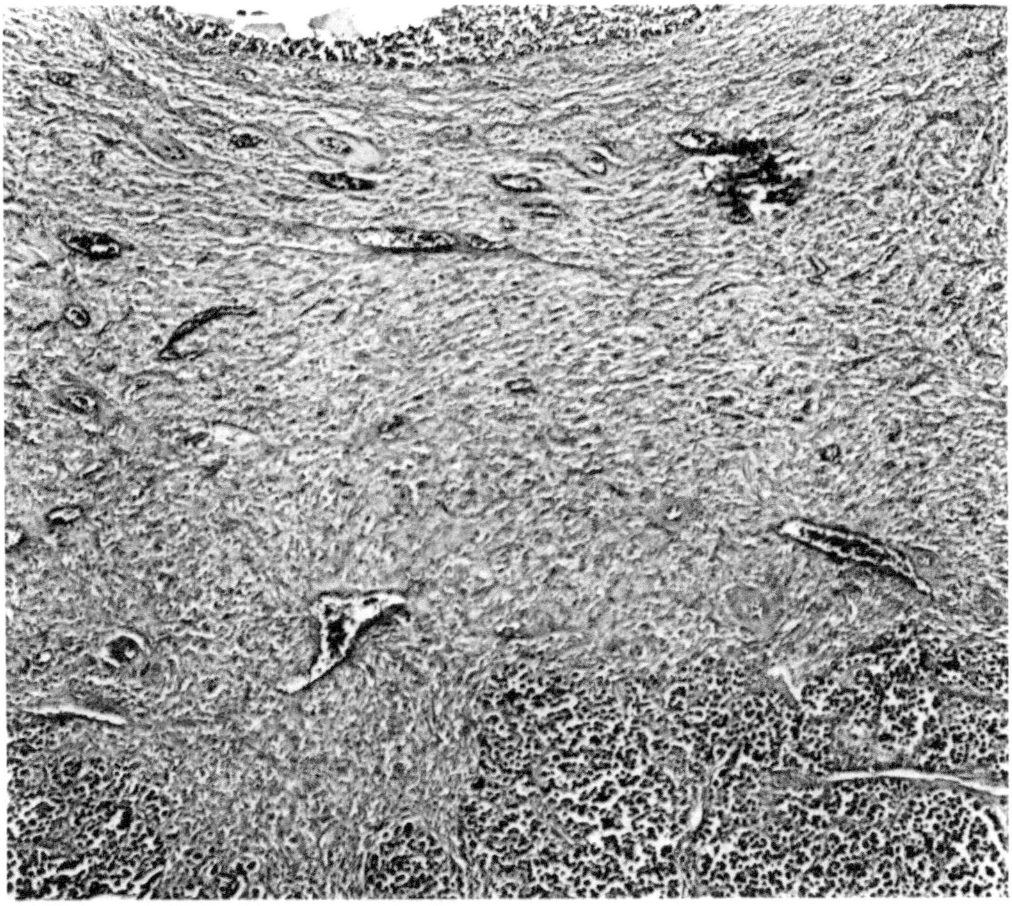

FIGURE 28. Section of ovary from an adult sea lion with a granulosa cell tumor invading the stroma of the ovary. There is a follicle at the upper edge, and the invading tumor cells across the bottom.

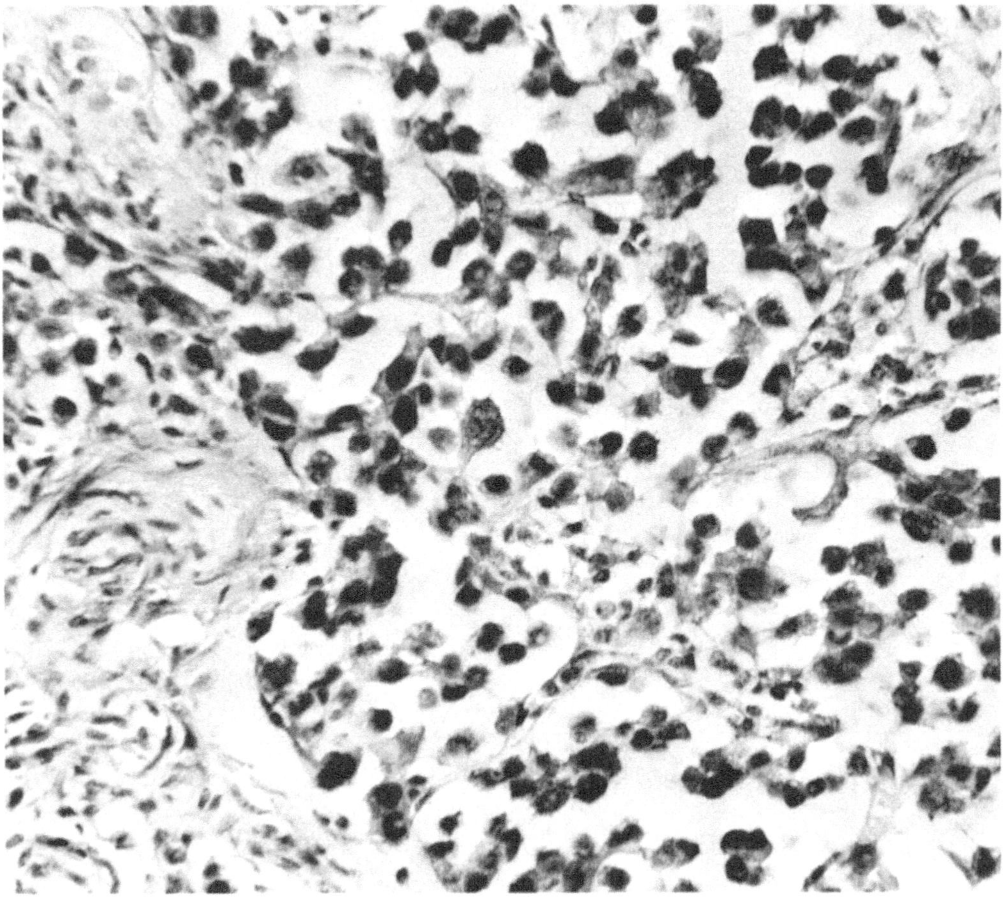

FIGURE 29. Higher magnification of granulosa cell tumor shown in Figure 28, showing the round hyperchromatic tumor cells arranged along a basement membrane, with minimal connective tissue stroma separating the clusters of tumor cells.

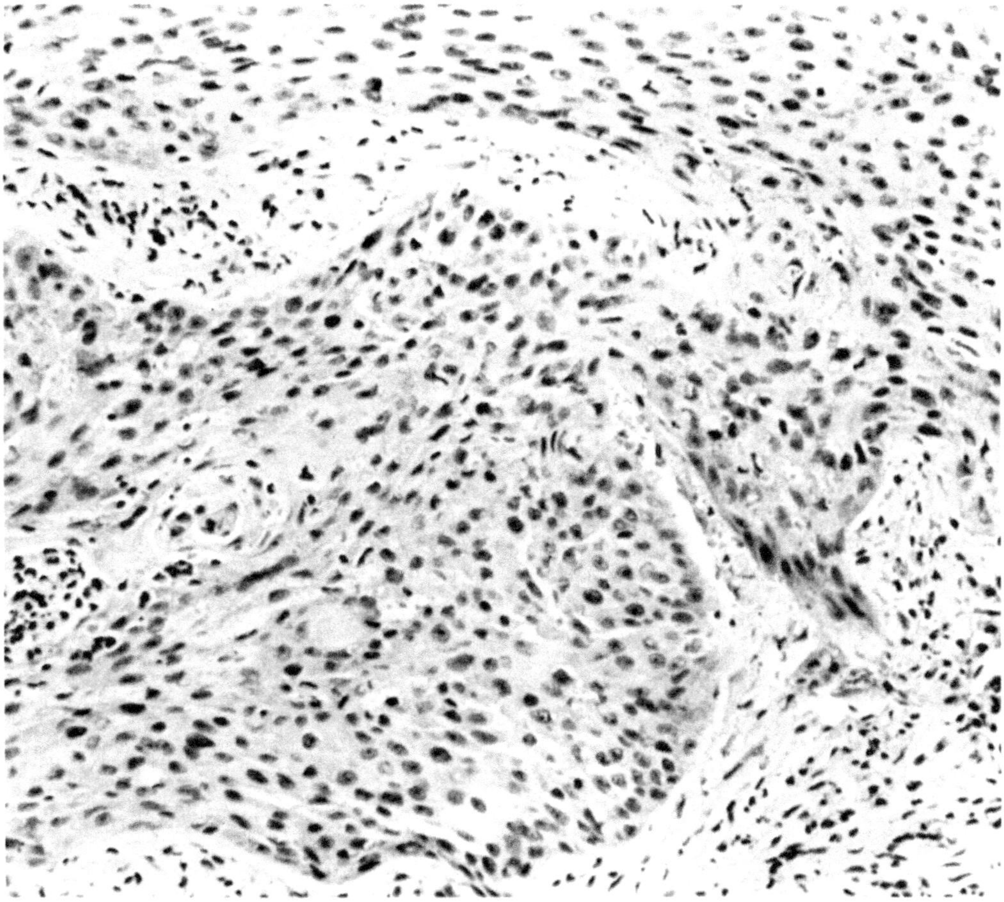

FIGURE 30. Tissue section from a squamous cell carcinoma of the cervix of an adult sea lion. Note the invading sheet of squamous epithelial cells, with evidence of early keratin pearl formation.

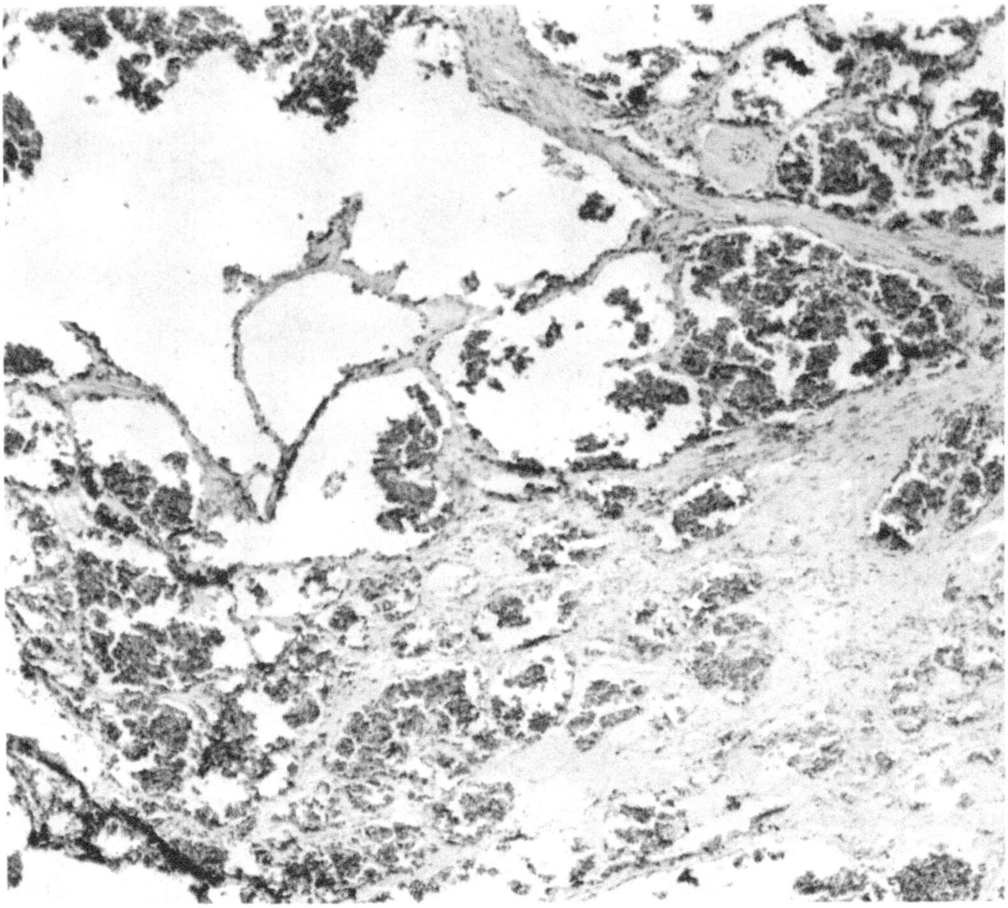

FIGURE 31. This is a section from a cavernous hemangioma found arising in the wall of the vagina of a sea lion, partly obliterating the cavity of that structure.

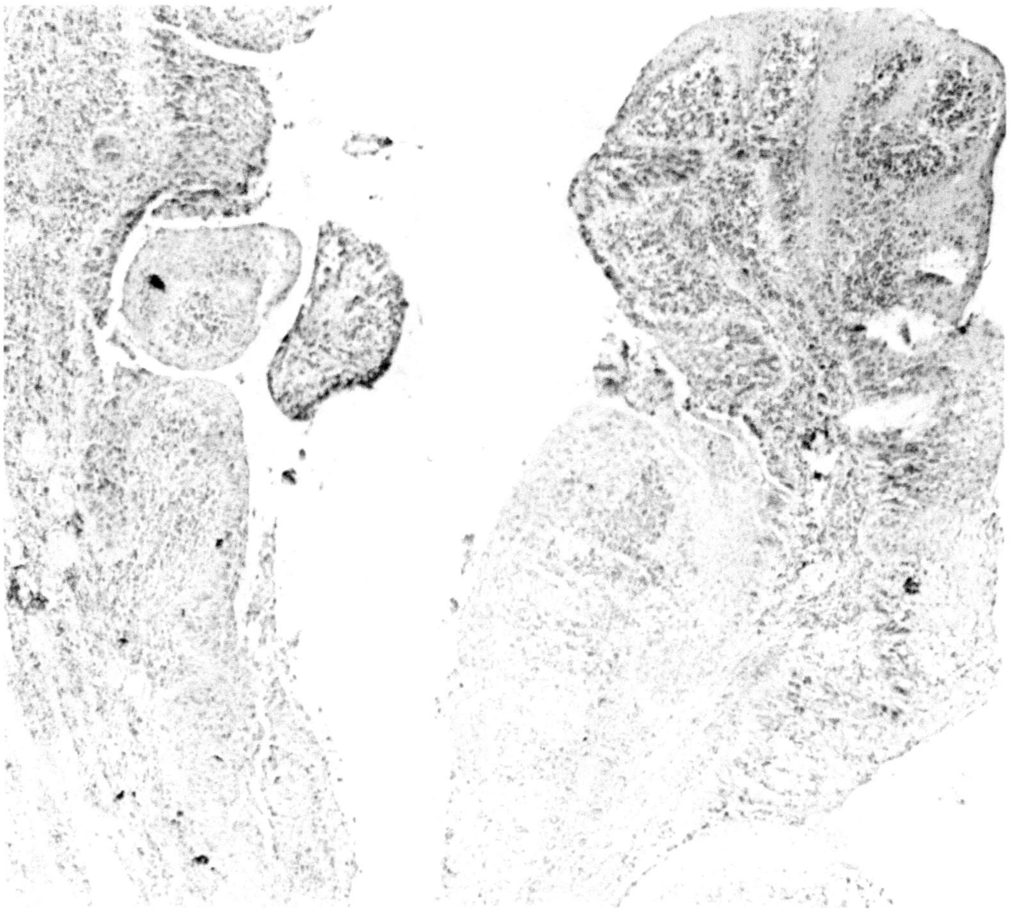

FIGURE 32. Tissue section showing one of the numerous papillomatous projections from the vaginal wall into the lumen, in a sea lion with multiple inflammatory papillomas of the vagina.

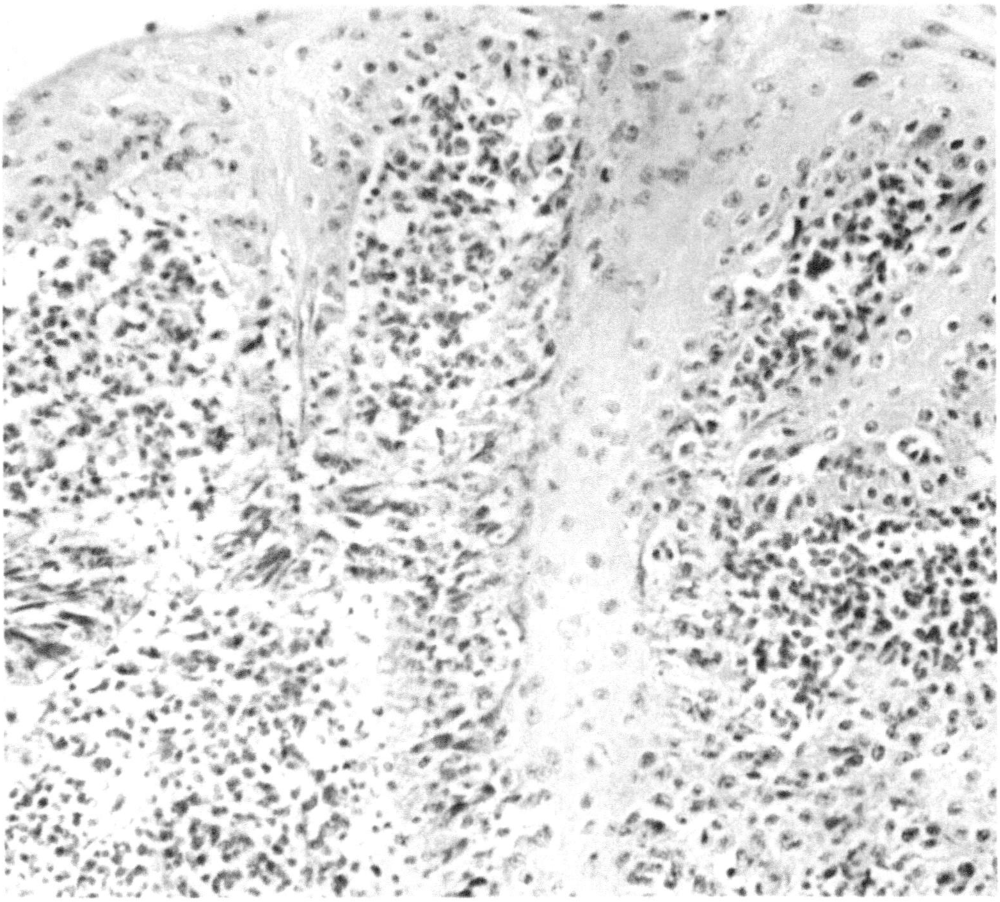

FIGURE 33. Higher magnification of the tip of a papillomatous growth from Figure 32, showing a dense stromal infiltration of lymphocytes and plasma cells.

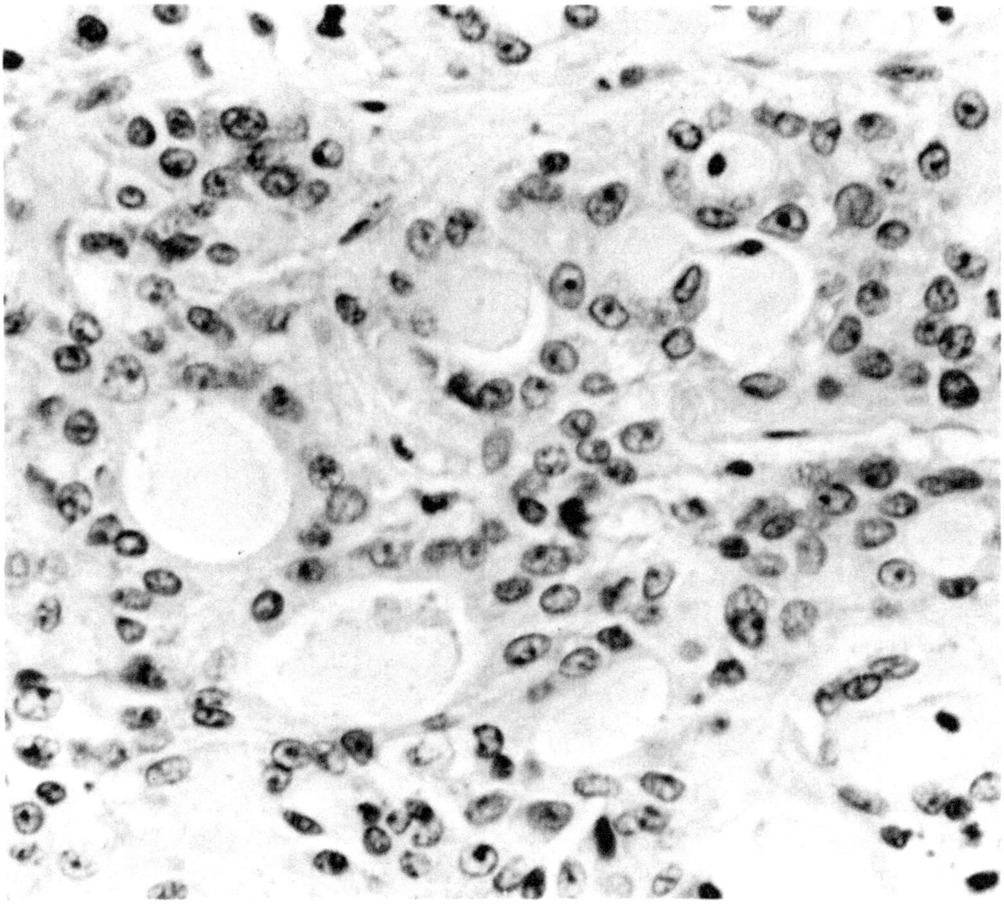

FIGURE 34. Section from a well differentiated acinar forming adenocarcinoma, apparently originating in the prostate gland of an aged sea lion.

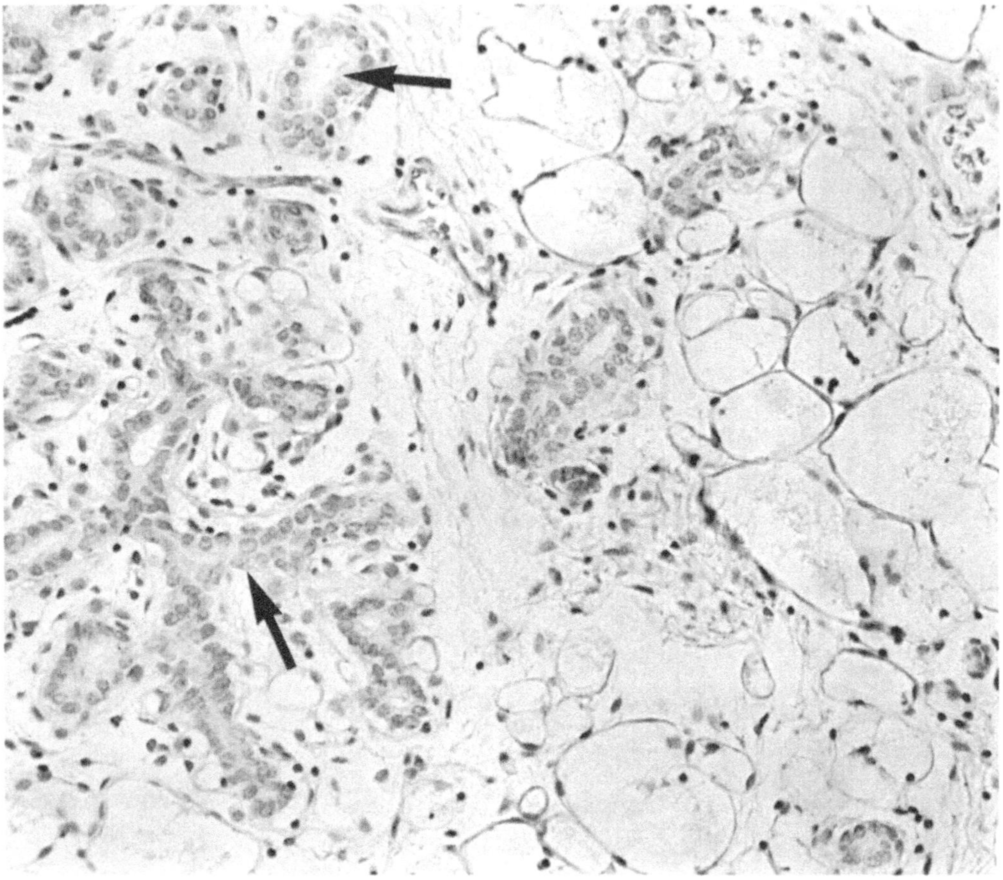

FIGURE 35. Section from a mammary gland mass of a sea lion, demonstrating a mixture of residual ductal elements (arrows), among an infiltrating mass of varying sized lipocytes.

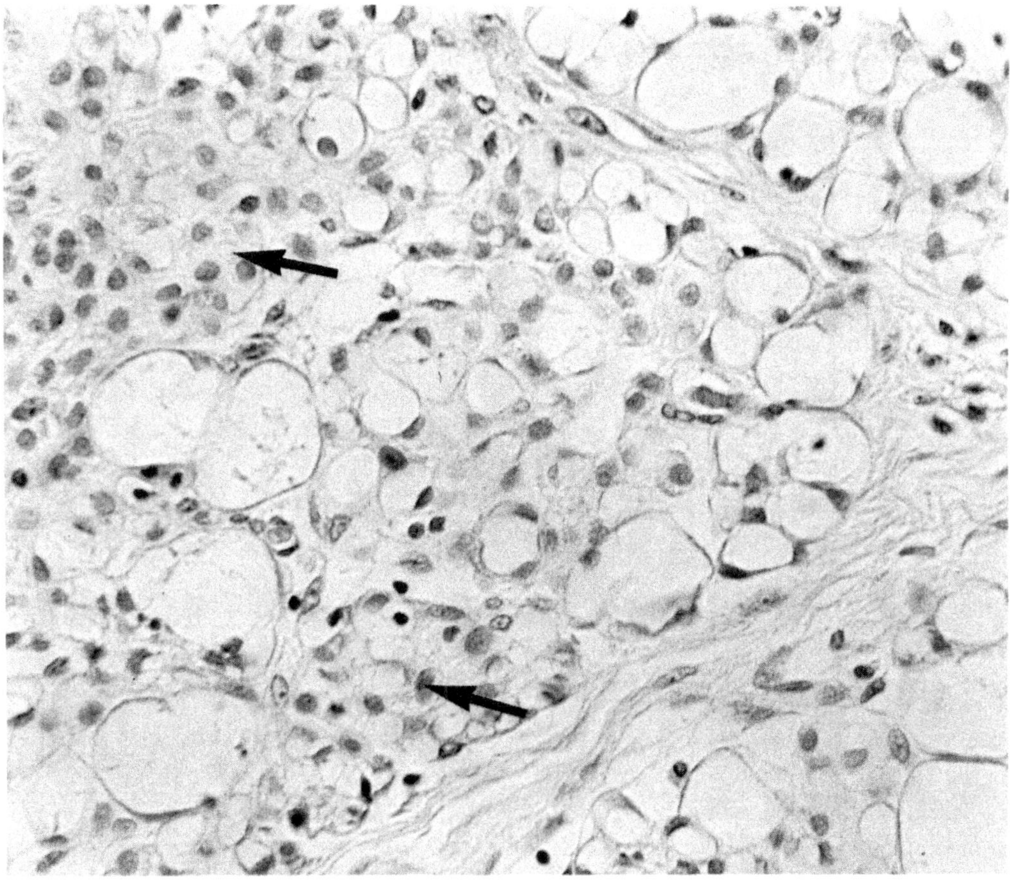

FIGURE 36. Higher magnification of mammary gland lipoma (Figure 35), showing undifferentiated (arrows) and differentiated lipoblasts and lipocytes.

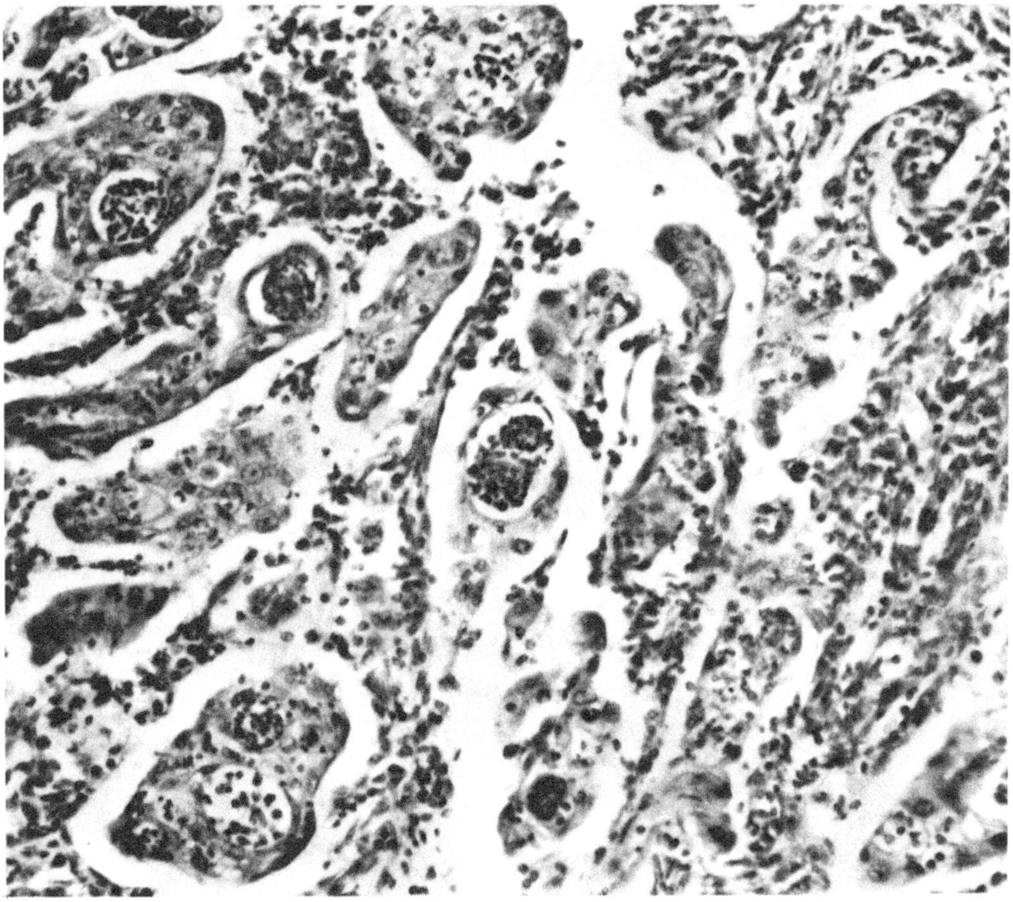

FIGURE 37. Section from a renal tubular adenocarcinoma found in an adult female California sea lion.

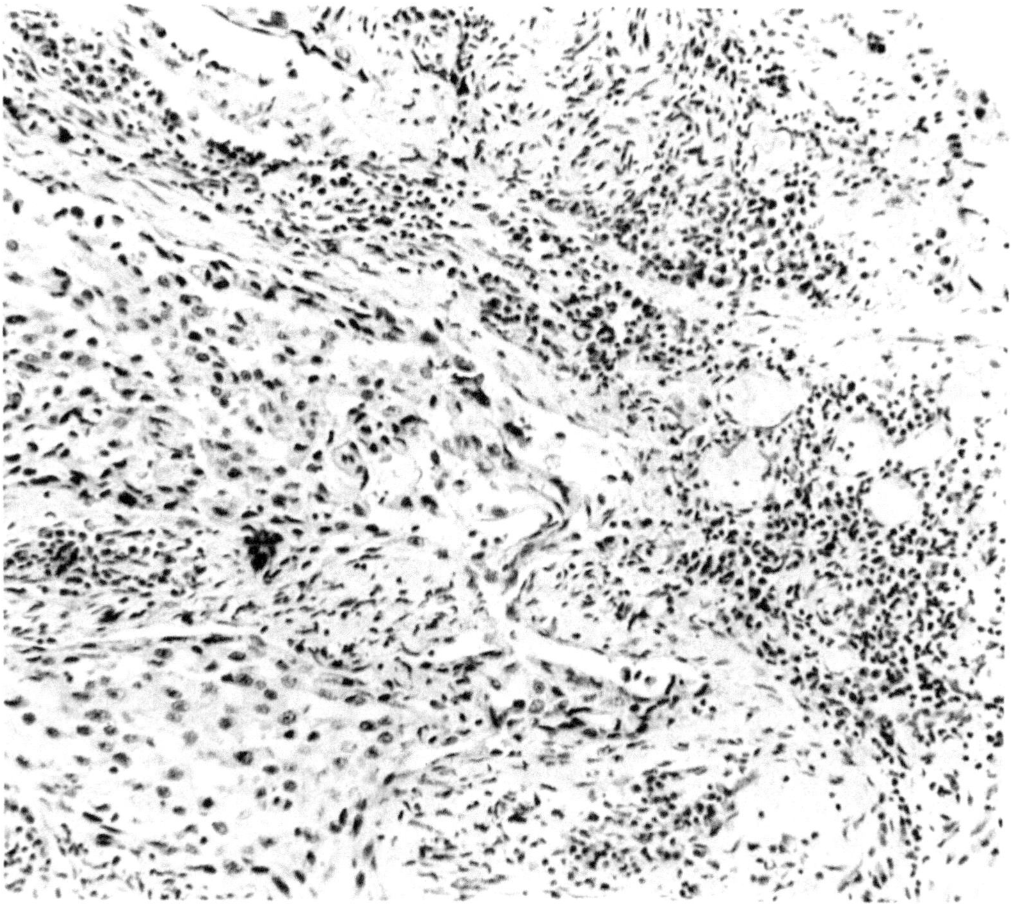

FIGURE 38. Metastatic focus of renal adenocarcinoma in a lymph node, demonstrating the marked squamous metaplasia of the tumor cells.

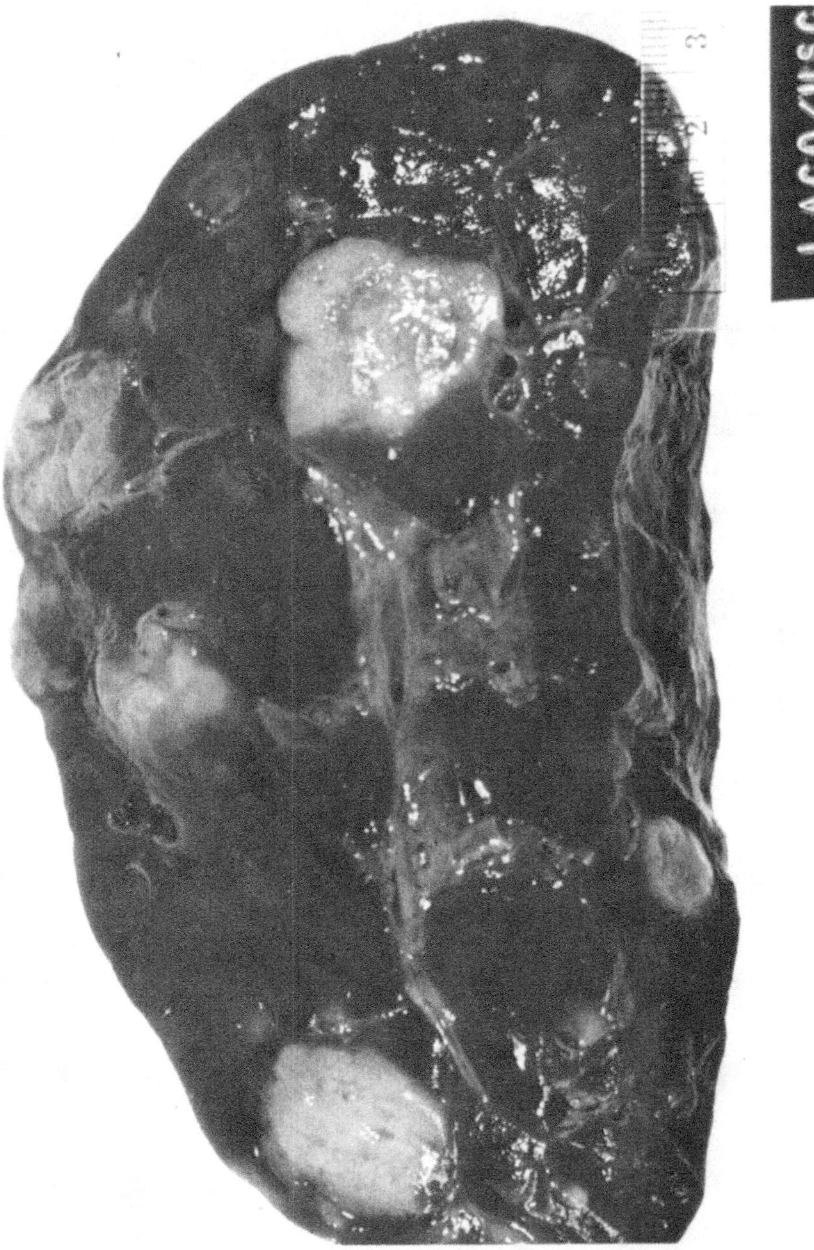

FIGURE 39. Gross photograph of renal adenocarcinoma from a mature sea lion. The neoplasm had multiple firm white nodules in both kidneys.

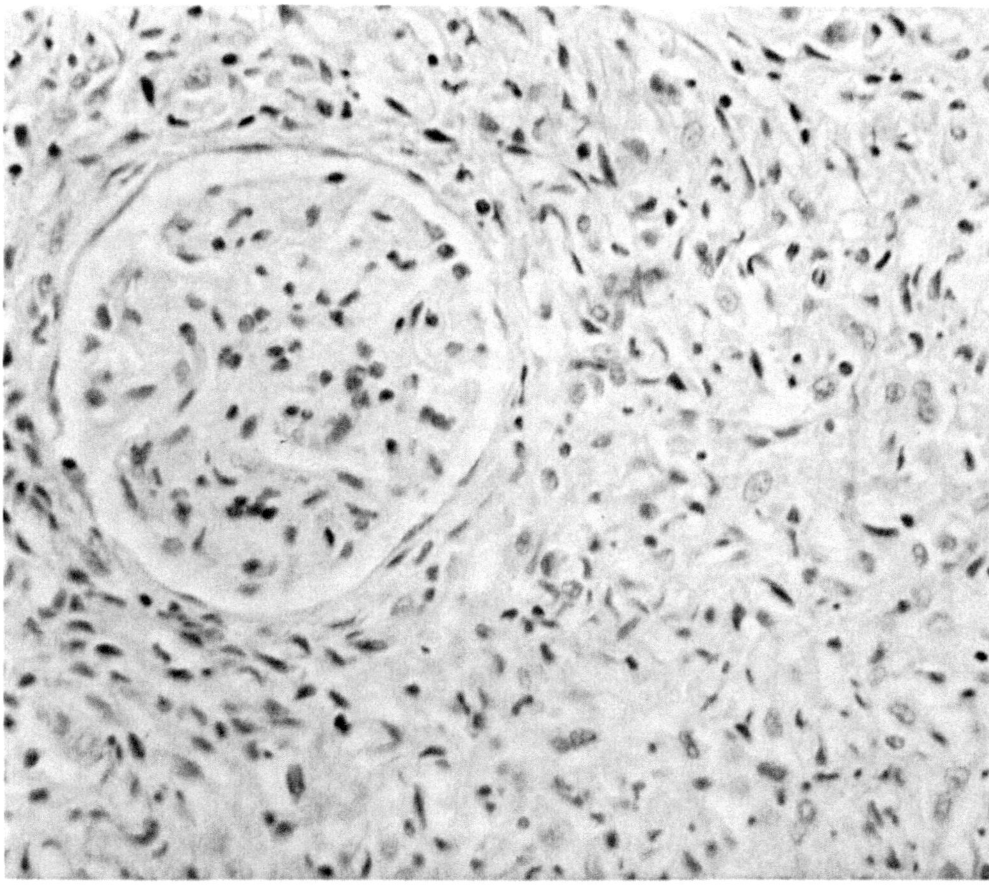

FIGURE 40. Section from the kidney shown in Figure 39, showing extensive cortical infiltration of anaplastic renal adenocarcinoma.

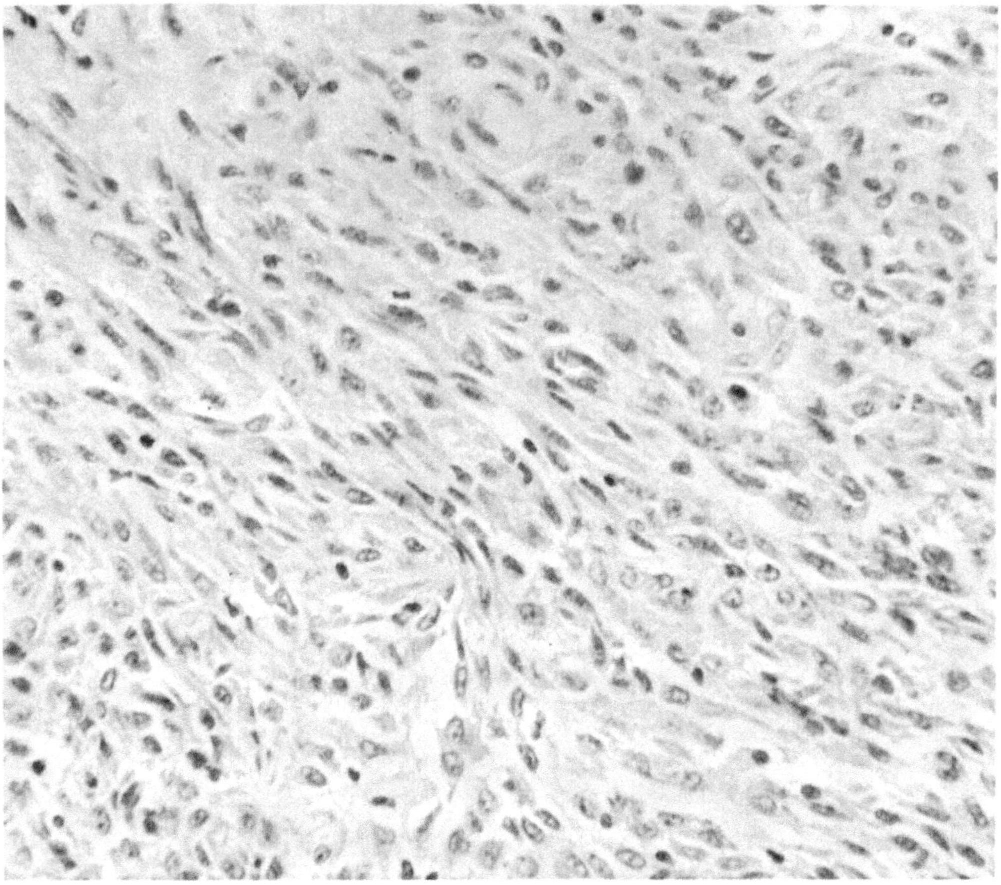

FIGURE 41. Cardiac metastasis from renal adenocarcinoma, showing marked anaplasia of tumor cells, with some spindle cell features.

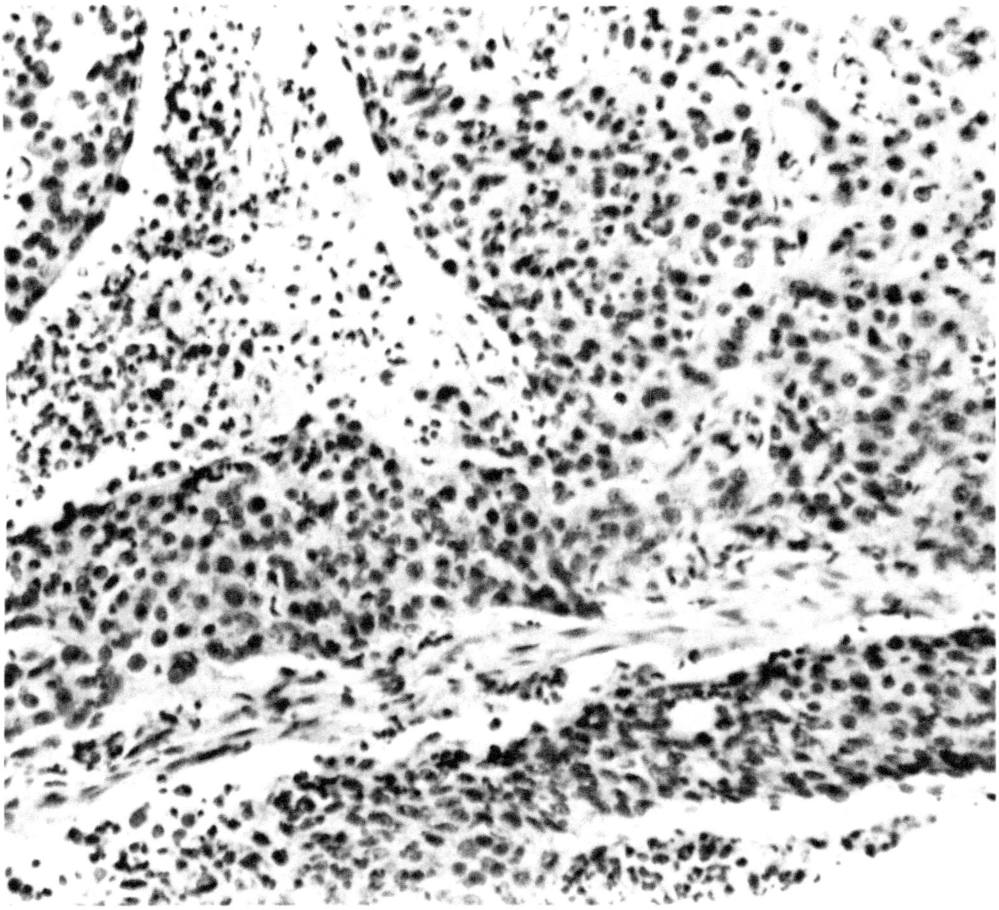

FIGURE 42. Tissue section from a transitional cell carcinoma of the urinary bladder of a sea lion. There is piling up of the tumor cells.

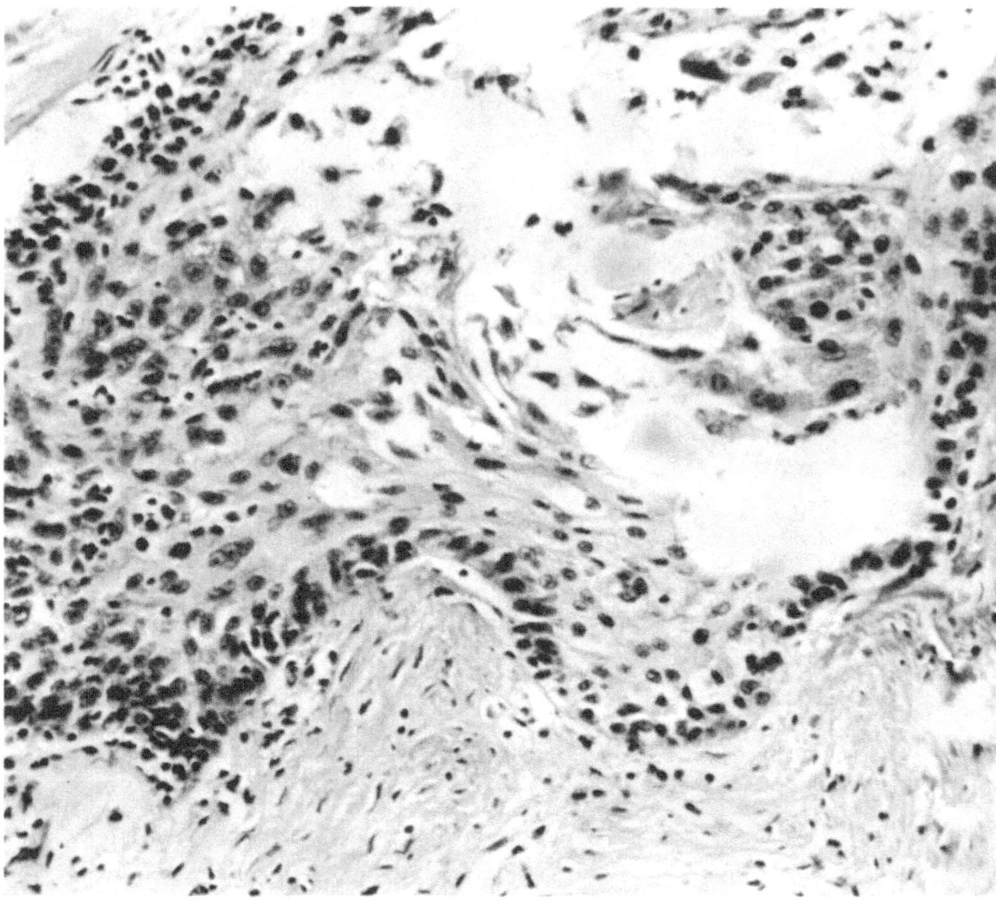

FIGURE 43. Metastasis from a transitional cell carcinoma, demonstrating prominent squamous features of the tumor cells.

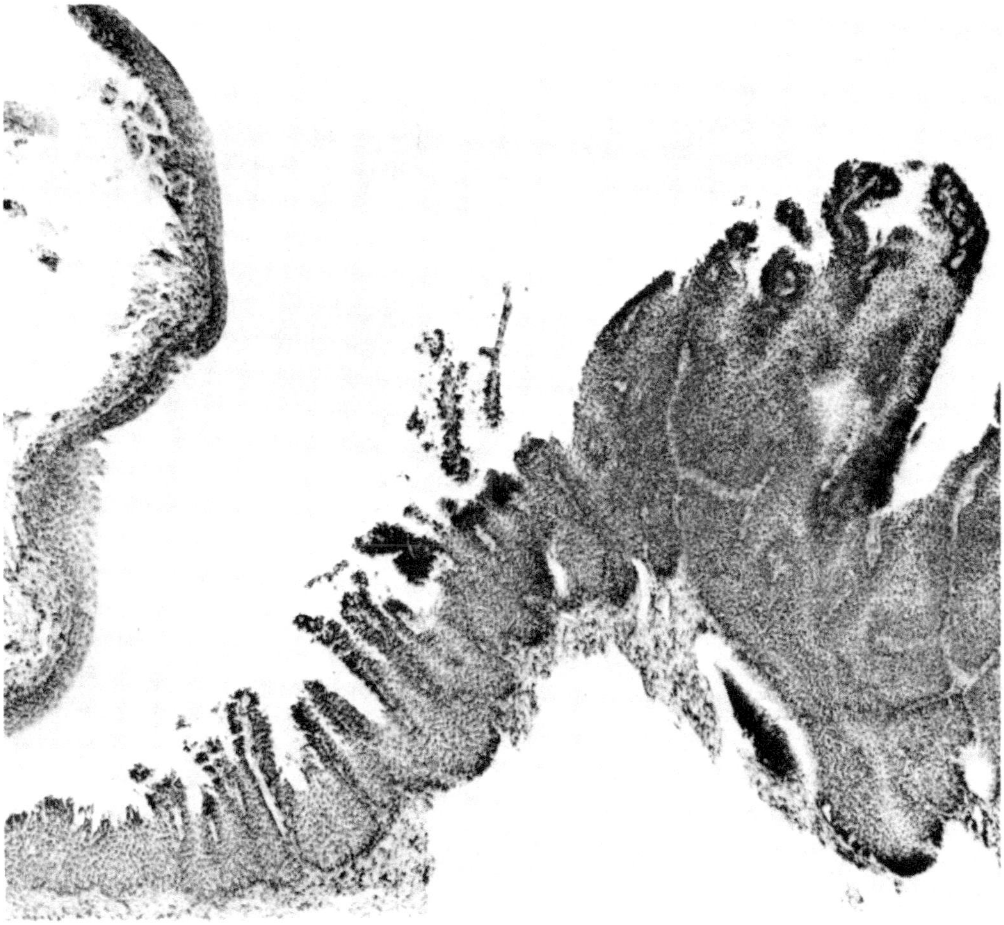

FIGURE 44. Bladder section from a sea lion, showing a papillomatous lesion projecting into the lumen of the bladder.

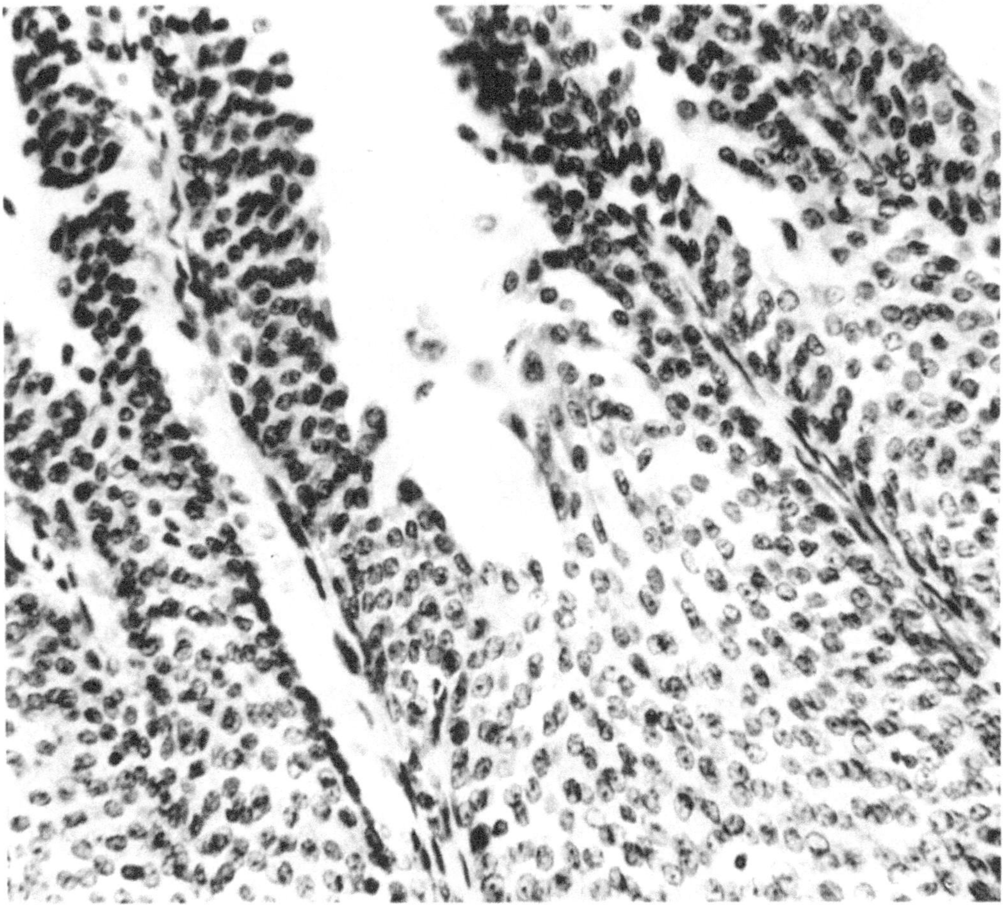

FIGURE 45. Higher magnification of bladder papilloma seen in Figure 44, showing well differentiated transitional epithelial cells forming finger like projections into the bladder lumen.

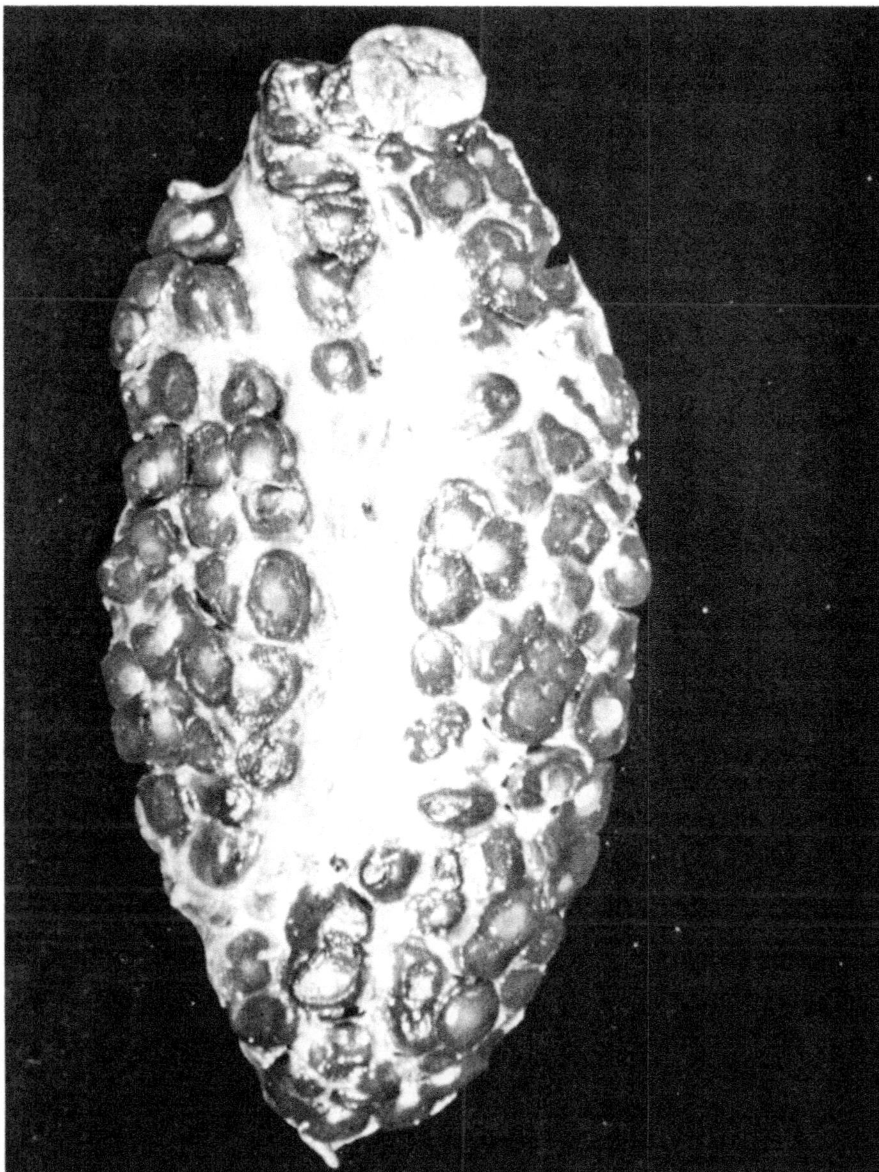

FIGURE 46. Nodular pale renal adenoma in an Atlantic bottle-nose dolphin. The lesion is at the right pole of the kidney. (Courtesy of Dr. G. Migaki and Comparative Pathology Registry, A.F.I.P).

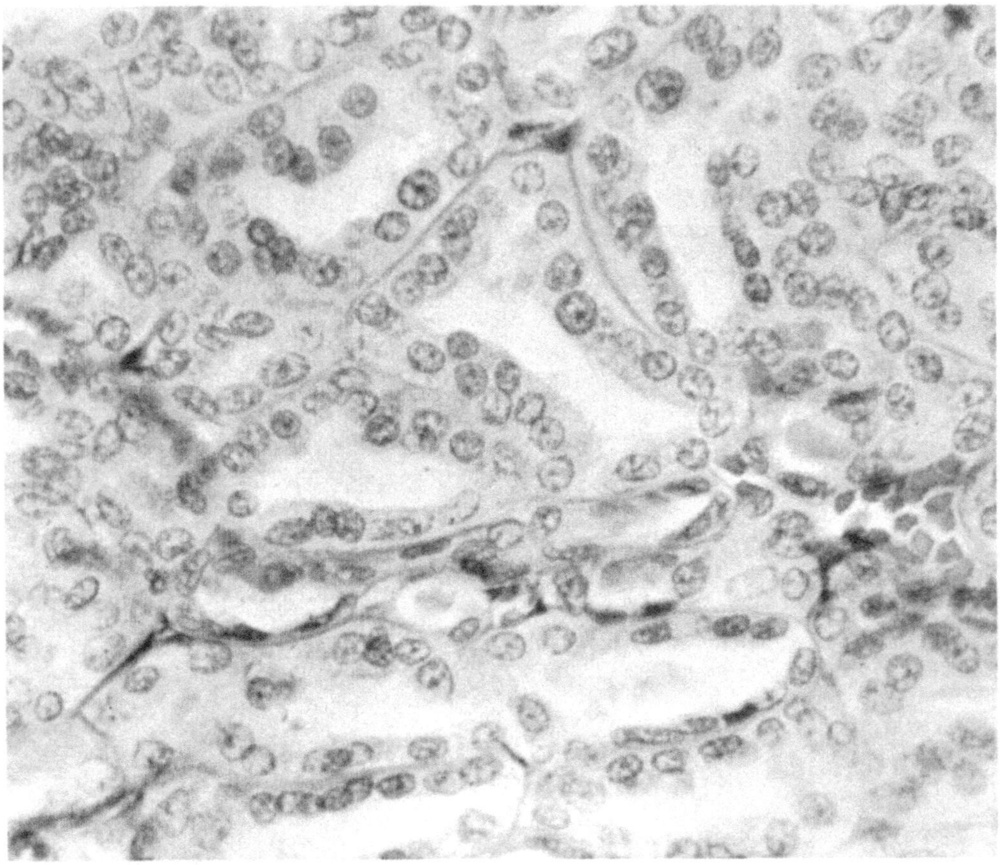

FIGURE 47. Tissue section from a well differentiated renal tubular adenoma, depicted in Figure 46. (Courtesy of Dr. G. Migaki and Comparative Pathology Registry, A.F.I.P.)

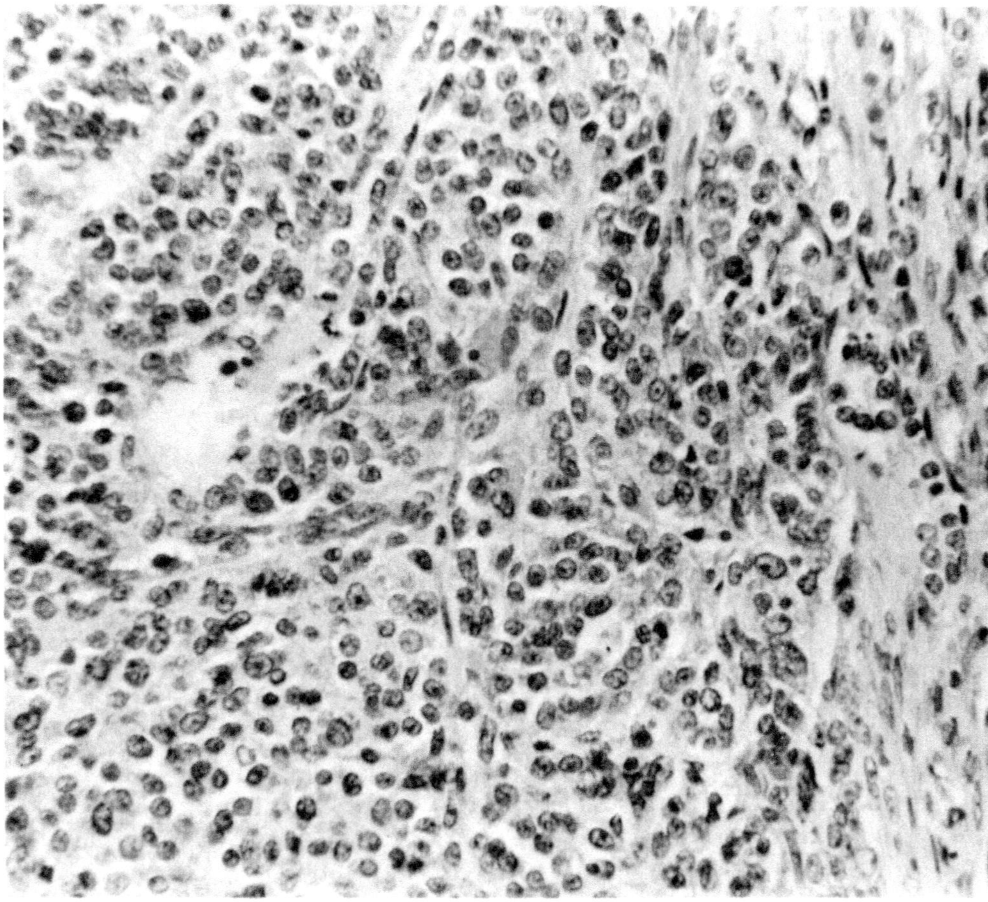

FIGURE 48. Tissue section of lymph node of a dolphin with malignant lymphoma. Note lymphoma cells, somewhat pleomorphic in appearance, at the left, and prominent fibrotic area along right side.

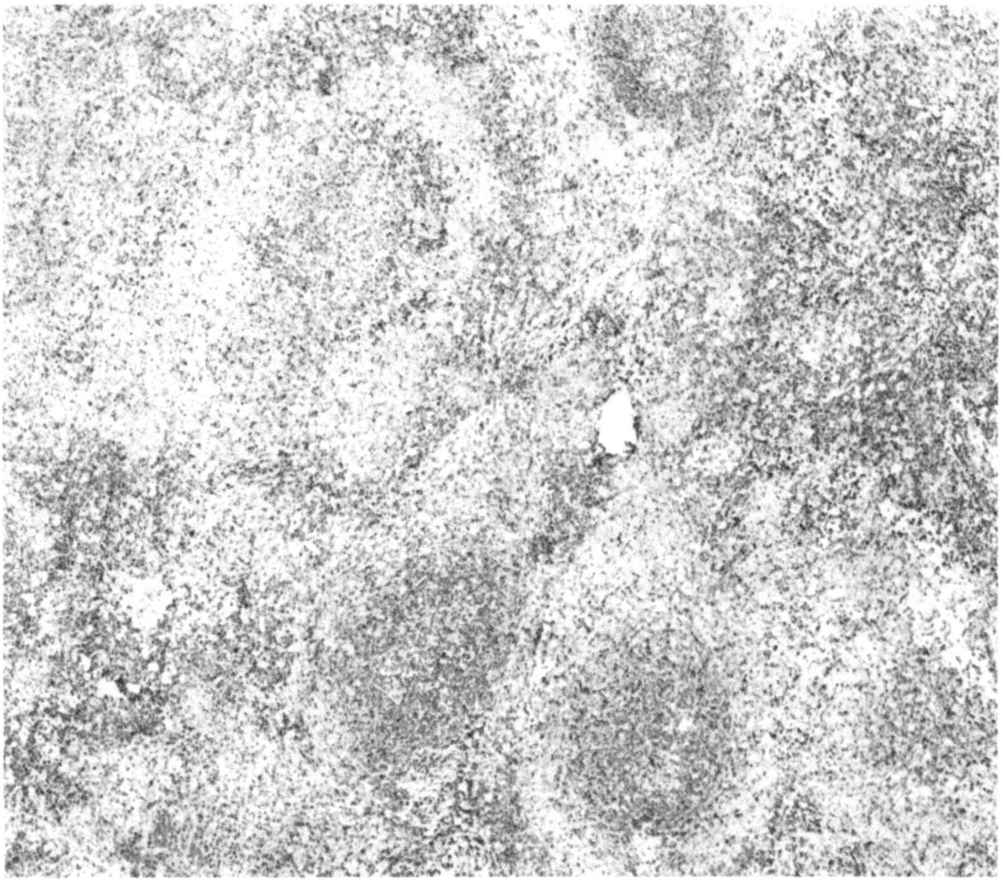

FIGURE 49. Section of spleen from a dolphin with eosinophilic leukemia. Note the hyperplastic lymphoid follicles, and extensive proliferative red pulp between the germinal follicles.

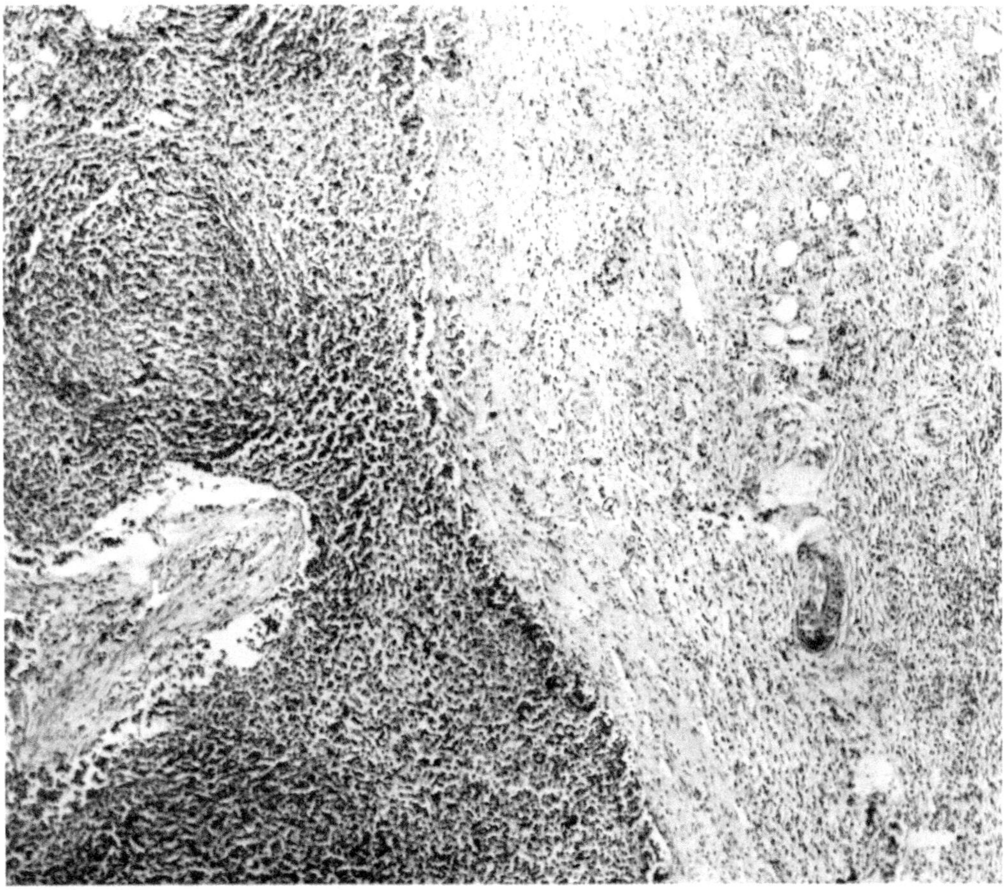

FIGURE 50. Section of mesenteric lymph node from the dolphin with eosinophilic leukemia, showing extensive thickening and fibrosis of capsular tissue, which is heavily infiltrated with mature and immature eosinophilic granulocytes.

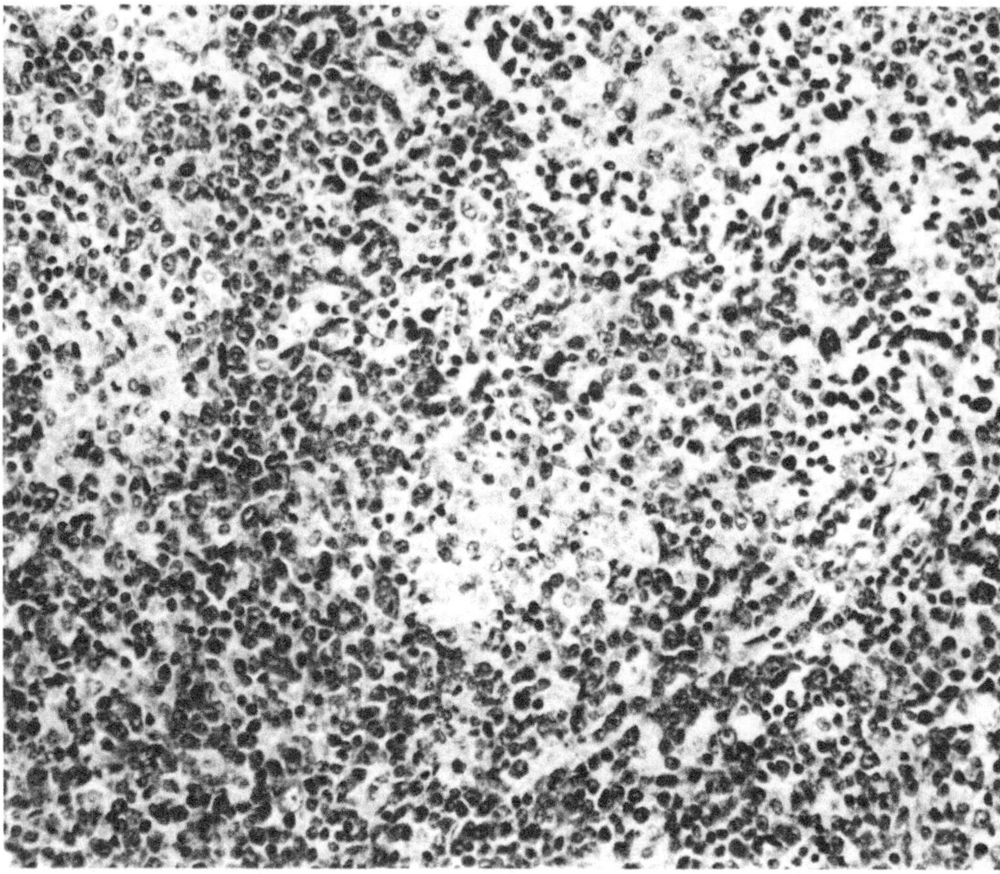

FIGURE 51. Section from the red pulp region of the spleen of the dolphin with eosinophilic leukemia. Note the extensive mitotic activity, and the predominance of blasts and other immature cells.

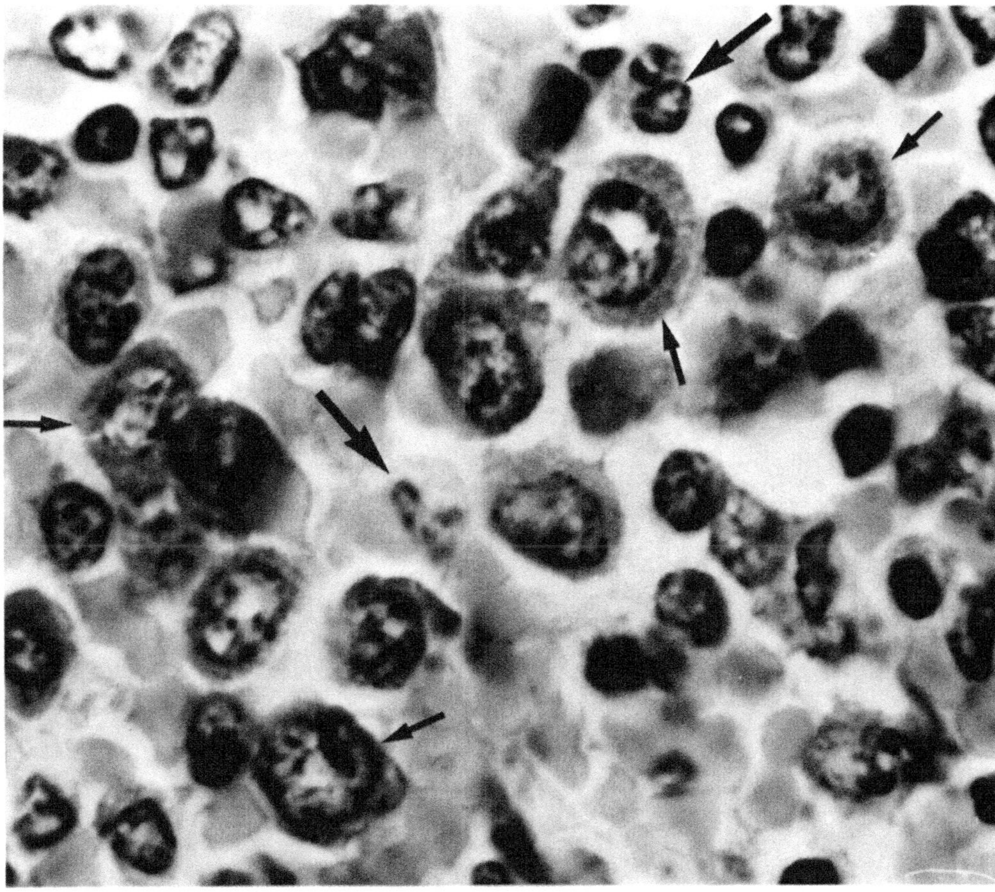

FIGURE 52. Higher magnification of the area of the spleen shown in Figure 51, demonstrating eosinophilic granulocytes at the promyelocyte and myelocyte stages of development (small arrows), as well as mature segmented cells (large arrows). Cytoplasmic granules are readily seen in these cells.

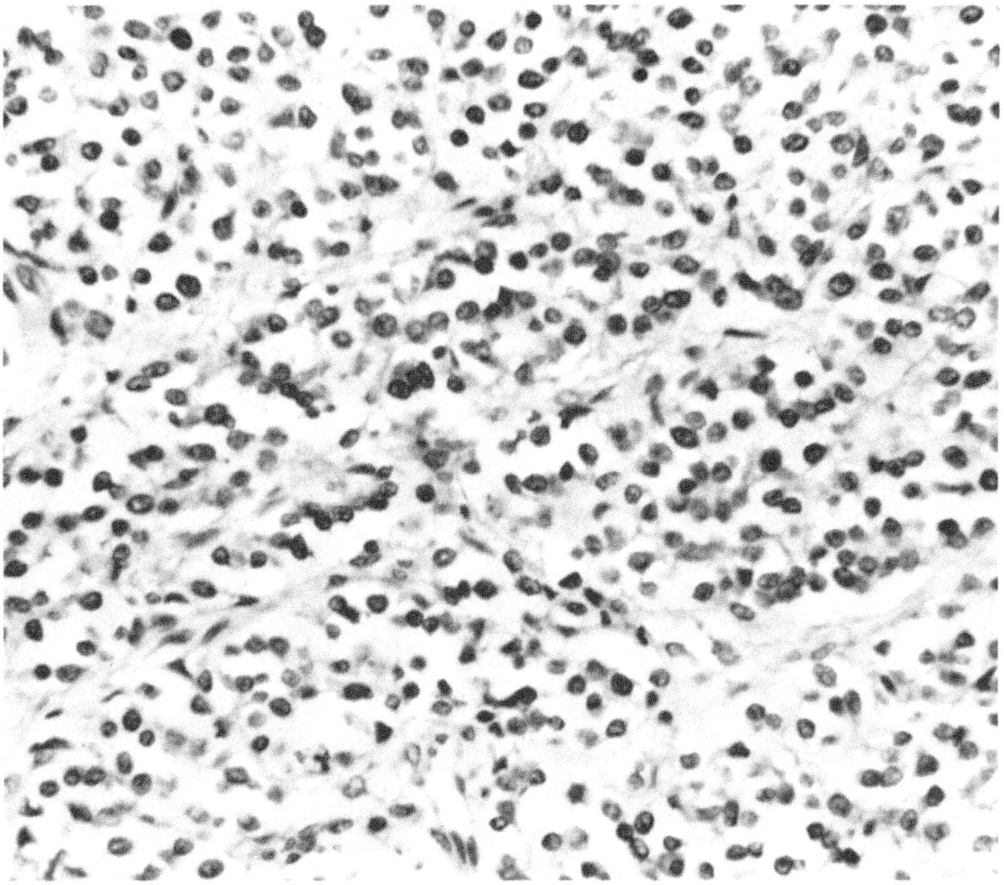

FIGURE 53. Tissue section from an abdominal mass found adjacent to and also involving the kidney of an adult sea lion, and the mass was classified as a neuroendocrine carcinoma (apudoma). Note the uniform appearance of the tumor cells, which have a large clear cytoplasm, and are arranged in a fine reticular stroma which gives an "organoid" appearance to the tumor, typical of tumors of this type.

REFERENCES

1. **Halloran, P. O.,** A bibliography of references to diseases in wild mammals and birds, *Am. J. Vet. Res.,* 16, 1, 1955.
2. **Fox, H.,** Adenocarcinoma and myosarcoma of mammary gland, *Rep. Penrose Res. Lab.,* 1941.
3. **Fox, H.,** *Disease in Captive Wild Mammals and Birds,* Lippincott, Philadelphia, 1923, 339.
4. Anonymous, 16th Annual Report, Washington Park Zoo, Milwaukee, Wisconsin, 1926.
5. **Timon-David, Jr.,** Sur une tumeur de l' estomac d' un dauphin *(Tursiops tursio), C. R. Soc. Biol.,* Paris, 107, 359, 1931.
6. **Timon-David, J.,** Estude sur une tumeur a trematodes de l'estomac d'un dauphin, *Bull. Inst.Oceanogr.,* Monaco, 1932, 11 pp.
7. **Rewell, R. E.,** Some tumors of wild animals, *J. Pathol. Bacterial.,* 62, 450, 1950.
8. **Rewell, R. E. and Willis, R. A.,** Some tumors found in whales, *J. Pathol. Bacterial.,* 61, 454, 1949.
9. **Stolk, A.,** Tumors in whales, *Amsterdam Naturalist,* 1, 28, 1950.
10. **Stolk, A.,** Some tumors in whales, in *Proc. K. Nederlandse Akademie van Wetenschappen,* Section C, 55, 215, 1952.
11. **Stolk, A.,** Some tumors in whales II, in *Proc. K. Nederlandse Akademie van Wetenschappen,* Section C, 56, 369, 1953.
12. **Stolk, A.,** Tumors in whales III, in *Proc. K. Nederlandse Akademie van Wetenschappen,* Section C, 65, 250, 1962.
13. **Simpson, J. G. and Gardner, M. B.,** Comparative microscopic anatomy of selected marine mammals, in *Mammals of the Sea, Biology and Medicine,* Ridgway, S. H., Ed., Charles C Thomas, Springfield, Ill., 1972, 298.
14. **Simpson, J. G.,** personal communication, 1982.
15. **Ridgway, S. H.,** Homeostasis in the aquatic environment, in *Mammals of the Sea, Biology and Medicine,* Ridgway, S. H., Ed., Charles C Thomas, Springfield, Ill., 1972, 590.
16. **Urp, C. J. and Best, P. B.,** Pathology of lesions observed in whales flensed at Saldahna Bay, South Africa, *J. Comp. Pathol.,* 76, 407, 1966.
17. **Cockrill, W. R.,** Pathology of the cetacea, Part 1, *Brit. Vet. J.,* 116, 133, 1960.
18. **Pilleri, G.,** Hirnlipom bein Buckelwal, *Megaptera novaenglial, Path. Vet.,* 3, 341, 1966.
19. **Landy, R. B.,** A review of neoplasia in marine mammals (*Pinnipedia and Cetacea*), in *The Comparative Pathology of Zoo Animals,* Montali, R. J. and Migaki, G., Eds., Smithsonian Institution Press, Washington, D.C., 1980, 579.
20. **Mawdesly-Thomas, L. E.,** Some aspects of neoplasia in marine animals, *Adv. Marine Biol.,* 12, 196, 1974.
21. **Cowan, D. F.,** Pathology of the pilot whale *Globicephala melaena, Arch. Pathol.,* 82, 187, 1966.
22. **Flom, J. O., Brown, R. J., Jones, R. E., and Schonewald, J.,** Vaginal fibromas in a beaked whale, *Mesoplodon densirostris, J. Wildlife Dis.,* 16, 99, 1980.
23. **Pilleri, G.,** Cerebrale neurofibroma beim Finnural, *Balaenoptera physalus, Pathol. Vet.,* 5, 35, 1968.
24. **Effrom, M., Griner, L., and Benirschke, K.,** Nature and rate of neoplasia found in captive wild mammals, birds, and reptiles at necropsy, *J. Natl. Cancer Inst.,* 59, 185, 1977.
25. **Stroud, R. K. and Roffe, T. J.,** Causes of death in marine mammals stranded along the Oregon coast, *J. Wildlife Dis.,* 15, 91, 1979.
26. **Brown, R. J., Smith, A. W., Morejohn, G. V., and DeLong, R. L.,** Metastatic adenocarcinoma in two California sea lions, *Zalophus c. californianum, J. Wildlife Dis.,* 16, 261, 1980.
27. **L. A. Griner, San Diego Zoo,** personal communication, 1981.
28. **Moore, J. J. and Stackhouse, L. J.,** Pancreatic duct adenoma and strangulation of the small intestine in a California sea lion *(Zalophus californianus), J. Wildlife Dis.,* 14, 445, 1978.
29. **Sweeney, J. C. and Gilmartin, W. G.,** Survey of diseases in free-living California sea lions, *J. Wildlife Dis.,* 10, 370, 1974.
30. **Nakajima, M. and Fujimoto, A.,** A case of retro-peritoneal tumor of a California sea lion (*Zalophus californianus*), compatible with malignant mesenchymoma, *Enoshima Marineland Reports,* 13, 1, 1971.
31. **Griner, L. A.,** Malignant leukemic lymphoma in two harbor seals, *Am. J. Vet. Res.,* 32, 827, 1971.
32. **Stroud, R. K. and Stevens, D. R.,** Lymphosarcoma in a harbor seal *(Phoca vitulina richardii), J. Wildlife Dis.,* 16, 267, 1980.
33. **Stedham, M. A., Casey, H. W., and Keyes, M. C.,** Lymphosarcoma in an infant northern fur seal *(Callorhinus ursinus), J. Wildlife Dis.,* 13, 176, 1977.
34. **Brown, R. J., Smith, A. W., and Keyes, M. C.,** Renal fibrosarcoma in the northern fur seal, *J. Wildlife Dis.,* 11, 23, 1975.
35. **Larsen, S.,** A survey of the postmortem findings in Pinnipedia autopsied during a 10-year period 1952—1961, *Nord. Vet. Med.,* 14, 150, 1962.

36. **Mawdesly-Thomas, L. E.,** An ovarian tumor in a southern elephant seal, *Vet. Pathol.*, 8, 9, 1971.
37. **Tierney, T. J.,** Disease and injury in the southern elephant seal, *Aust. Vet. J.*, 53, 91, 1977.
38. **Mawdesley-Thomas, L. E. and Bonner, W. N.,** Uterine tumors in a grey seal, *J. Pathol.*, 103, 205, 1971.
39. **Williams, T. D. and Pulley, L. T.,** Leiomyomas in two sea otters, *Enhydra lutris, J. Wildlife Dis.*, 17, 401, 1981.
40. **Ridgway, S. H.,** Reported causes of death of captive killer whales *(Orcinus orca), J. Wildlife Dis.*, 15, 99, 1979.
41. **Migaki, G.,** Renal adenoma in an Atlantic bottle-nosed dolphin *(Tursiops truncatus), Am. J. Vet. Res.*, 39, 1920, 1978.
42. **Bengtsson, E.,** Eosinophilic leukemia — an immunopathologic reaction? *Acta Paediat. Scand.*, 57, 245, 1968.
43. **Pierce, L. E., Hasseinran, A. H., and Constantine, A. B.,** Disseminated eosinophilic collagen disease, *Blood*, 29, 540, 1967.
44. **Meissner, W. A. and Diamandopoulos, G. Th.,** Neoplasia, in *Pathology*, 7th ed., Anderson W. A. D. and Kissane, J. M., Eds., C. V. Mosby, St. Louis, 1977, 688.
45. **Stunzi, H., Head, K. W., and Nielson, S. W.,** Tumours of the lung, *Bull. W.H.O.*, 50, 9, 1974.
46. **Howard, E. B., Esra, G. N., and Young, D.,** Acute foodborne pesticide toxicity in cormorants and seagulls, in *Animals as Monitors of Environmental Pollutants*, National Academy of Sciences, Washington, D.C., 1979, 290.
47. **Young, D. R., Heesen, T. C., Esra, G. N., and Howard, E. B.,** DDE-contaminated fish off Los Angeles are suspected cause in deaths of captive marine birds, *Bull. Environ. Contamin. Toxicol.*, 21, 54, 1979.
48. **Young, D. R. and Mearns, A. J.,**Pollutant flow through food webs, in *Southern California Coastal Water Research Project*, Bascom W., Ed., Southern California Coastal Research Project, Long Beach, Calif., 1978, 185.
49. **Delong, R., Gilmartin, W. G., and Simpson, J. G.,** Premature births in California sea lions: associated with high organochlorine pollutant residue levels, *Science*, 181, 1168, 1973.
50. **Mearns, A. J. and Sherwood, M. J.,** Distribution of neoplasms and other diseases in marine fishes relative to the discharge of waste water, in *Aquatic Pollutants and Biologic Effects with Emphasis on Neoplasia*, Kraybill, H. F., Dawe, C. J., Harshbarger, J. C., and Tardiff, R. G., Eds., New York Academy Science, 298, 1977, 210.
51. **Puffer, H., Duncan, K., von Hofe, E., Winkler, D., Brewer, G., Mondal, S.,** Benzo(a)pyrene: studies on the biologic effects of this ubiquitous pollutant on fishes, *Ocean 79*, Institute of Electrical and Electronics Engineers, Inc., N.Y., 7, 398, 1979.
52. **Mix, M. C., Trenholm, S. R., and King, K. I.,** Benzo(a)pyrene body burdens and the prevalence of proliferative disorders in mussels (*mytilus edulis*) in Oregon, in *Animals as Monitors of Environmental Pollutants*, National Academy of Sciences, Washington, D.C., 1979, 52.
53. **Stich, H. F., Acton, A. B., Oishi, K., Yamazaki, F., Harada, T., Hibino, T., and Moser, H. G.,** Systematic collaborative studies on neoplasms in marine animals as related to the environment, in *Aquatic Pollutants and Biologic Effects with Emphasis on Neoplasia*, Kraybill, H. F., Dawe, C. J., Harshbarger, J. C., and Tardiff, R. G., Eds., New York Academy Science, 298, 1977, 374.
54. **Stetzer, E., Williams, T. D., and Nightingale, J. W.,** Cholangiocellular adenocarcinoma, leiomyoma, and pheochromocytoma in a sea otter, *J.A.V.M.A.*, 179, 1283, 1981.
55. **Zachary, J. F. and McDonald, S. C.,** Malignant teratoma in a spot-neck otter, *J.A.V.M.A.*, 179, 1279, 1981.
56. **Nelson, J. A., Struck, R. F., and James, R.,** Estrogenic activities of chlorinated hydrocarbons, *J. Toxicol. Environ. Health*, 4, 325, 1978.
57. **Fuller, G. B., Knauf, V., Mueller, W., and Hobson, W. C.,** PCB augments LH-induced progesterone synthesis, *Bull. Environ. Contamin. Toxicol.*, 25, 65, 1980.
58. **Freeman, H. C., Uthe, J. F., and Sangalang, G.,** The use of steroid hormone metabolism studies in assessing the sublethal effects of marine pollution, *Rapp. P.-V. Reun. Cons. Int. Explor. Mes.*, 179, 16, 1980.
59. **Dorman, B. H., Genta, V. M., Mass, M. J., and Kaufman, D. G.,** Benzo(a) pyrene binding to DNA in organ cultures of human endometrium, *Cancer Res.*, 41, 2718, 1981.
60. **Migaki, G. and Albert, T. F.,** Lipoma of the liver in a bowhead whale *(Balaena mysticetus), Vet. Pathol.*, 19, 329, 1982.

Chapter 6

MISCELLANEOUS DISEASES

Edwin B. Howard

TABLE OF CONTENTS

I. MISCELLANEOUS DISEASES

It is not the intent of this chapter to attempt an exhaustive description of all of the many miscellaneous disease conditions, or undetermined etiology or pathogenesis, which are found with some frequency in marine mammals. It is more the purpose of this section to describe some diseases or lesions which are seen with some frequency, or are of a particular interest for comparative pathologic study, or perhaps of some conditions which we feel warrants description and further study. Many of the diseases brought together here have been described as individual case reports or part of surveys of marine mammal diseases in various scientific journals.

It becomes so obvious when we start to describe some the lesions in animals in which we don't even know the normal functional physiology of a particular organ, and in which a specific etiology has not been determined, that we must resort to drawing comparisons with similar lesions in other animal species or man in which the mechanism is understood. This is often a productive intellectual experience, and the conclusions may in fact be correct. However, one must not accept these speculations as fact, but should be aware that extensive research remains to be done to more accurately define the disease processes that we don't completely understand. This can be a formidable task when working with modest budgets, limited laboratory facilities, and a near lack of specific diagnostic reagents applicable to this group of animals.

Some of the published records of certain lesions in marine mammals are of questionable validity, because the observer was unaware of some of the rather unusual normal anatomic features present in these animals which may have inadvertently been described as a pathologic process.

The excellent review by Simpson and Gardner[1] shows much of the normal microscopic anatomy of certain marine mammals, but also describes many incidental pathologic processes encountered in their study, some of which are rather common, while others are rare.

As far as is feasible, the lesions described in this chapter will be presented according to organ system, realizing that in some cases the lesions have multi-organ distribution or effects, and we are usually describing the most obvious or most severe lesion. Some of the described lesions are incidental findings during necropsy examinations of animals that died of other obvious disease, and this points out the desirability of doing a thorough necropsy examination if at all possible. In many cases, the lesions described in this chapter were the primary disease and cause of death.

II. INTEGUMENT

Bacterial, viral, and parasitic lesions affecting the skin and adnexa have been discussed under those relevant chapter headings.

Traumatic injury to the skin is a common finding in both feral and captive marine mammals. Skin trauma, with erosion of the epidermis, is commonly seen in captive cetaceans, from chronic rubbing on the wall of the pool while swimming.

We have seen both dolphins and sea lions with lacerations which were apparently made by the propeller of motor boats. It is not uncommon for boaters to chase these animals. In several instances we have seen severe trauma in addition to skin lacerations. A skull fracture was found in one sea lion which had skin lacerations, and it was believed that this occurred from impact with a boat. Another young sea lion had a compound fracture of the flipper, which was of some duration as indicated by the extensive necrosis, abscessation, and osteomyelitis.

Severe skin lacerations caused by shark attacks are occasionally seen, and are easy to recognize because of the narrow deep parallel lacerations made by the shark teeth (Figure

1). When great white sharks have been caught along our coastal area, it is not uncommon to find a whole young sea lion in the stomach of the shark.

In our series of marine mammals, we have had several that died of gunshot wounds. It is likely that some are shot by commercial fishermen, but others are probably shot along the shore by perverted individuals exhibiting their marksmanship. In a study of the causes of death of marine mammals along the Oregon coast, 30% were found to have died of gunshot wounds.[2] Sometimes the animals are killed with high powered rifle fire, others are shot with hand guns, and in some cases shotguns have been used and the pellets are still found in the dermis or subcutis (Figure 2).

Endocrine disorders affecting the skin have not been reported in marine mammals. Very little is known of the normal functional status of the endocrine system of these animals, only sketchy information is available as to normal blood levels of various hormones. We have run thyroid and cortisol levels on several California sea lions, using RIA, and in those cases there appeared to be a normal T_3 and low T_4 level, and the cortisol appeared similar to other mammals. We do not know if the technique was reliable for this species of animal. Ridgway has carefully studied thyroid function in the dolphin,[3] but data on the sea lion is scanty. We have had one adult California sea lion which exhibited alopecic skin lesions which on histologic evaluation were quite suggestive of a hypothyroid condition. The skin sections showed thinning of the epidermis, considerable hyperkeratosis, and marked dilation of the cystic, keratin filled hair follicles (Figure 3), lesions which are quite reminiscent of hypothyroid skin disease in other animal species.

III. NERVOUS SYSTEM

The most common neuropathologic disease entity in our studies has been the parasitic encephalopathy so frequently observed in stranded dolphins, caused by the fluke of the *Nasitrema* sp. This is described under parasitic diseases.

We have observed, in two dolphins and one sea lion of our series, hemorrhagic encephalopathy which appeared grossly and microscopically identical to cerebrovascular occlusion and infarction as seen in other mammals, particularly man. In one dolphin, there was extensive subdural "paintbrush" hemorrhage over the left cerebral hemisphere (Figure 4). On cutting the brain, there was a 1 cm diameter area of acute hemorrhage and necrosis, fairly superficial in the anterior lateral cerebral hemisphere, with some hemorrhage and necrosis extending outward from the central infarcted core (Figure 5). On microscopic examination of tissue sections from this area, there was considerable subdural hemorrhage over the surface of the cerebral hemisphere (Figure 6), as well as extensive hemorrhage in the brain parenchyma at the site of the infarct (Figure 7). In the tissue adjacent to the necrotic core, there was acute ischemic necrosis and early collapse of the neuropil, making the existing small blood vessels much closer together, giving the appearance of greater vascularity. This type of lesion must be differentiated from a pre-existing capillary vascular anomaly, which often predisposes to a hemorrhagic infarct. In the latter lesion, however, there is not the abundance of brain parenchyma separating the capillary spaces, such as was found in the brains of the two dolphins with acute infarcts. Interestingly, the cerebrovascular infarcts in the two dolphins were similarly located, superficially in the cerebral hemisphere. The second dolphin did not show the marked subdural hemorrhage that was present in the first case.

The sea lion was a former stranded animal that had been in captivity for a year. Over a period of a week, it had become progressively more uncoordinated and ataxic, with a head tilt, and it was unable to climb on a platform. Gross and microscopic examination of the brain, after the animal had died, revealed that there were numerous areas of infarction, with malacia of varying degrees of duration, in the internal capsule, putamen, mid-brain, and brain stem. The sections showed no inflammation, merely areas of liquefaction necrosis and cavitation (Figure 8), with marked gitter cell proliferation to phagocytize the necrotic brain tissue. The animal was found to have some bronchopneumonia, probably terminal, and no etiology was found for the brain lesions.

We had one stranded elephant seal in our series that had typical microscopic lesions which were quite suggestive of lead poisoning. Since it was a feral animal, we had no positive evidence of lead consumption. It has been shown that lead levels in Los Angeles River run off water have declined over five-fold over the past decade, from levels of 940 $\mu g/\ell$ to 160 $\mu g/\ell$, due to decreased used of lead additive in gasoline.[4] The lesions which are detected in this elephant seal included widespread degenerative changes in neurons in the brain, so-called "red" neurons with hyperchromatic and sometime pyknotic nuclei (Figure 9), and marked satellitosis of microglial cells and frank neuronophagia. In addition to the lesions of the brain, there was also extensive toxic tubular nephrosis in the convoluted tubules of the renal cortices (Figure 10). These brain and renal changes have been described as associated with chronic lead poisoning in other species.

One problem that is commonly encountered in pinnipeds, when adapting them to a captive environment, is the development of corneal opacities (Figure 11). This has been thought to be caused by placing the animals in freshwater pools, but that is not a likely explanation since it is also seen in animals such as the one shown in Figure 11, that are placed in a circulating ocean water source pool such as those at Marineland of the Pacific. It may be from chemical contamination or superficial infection of the cornea, but the exact cause remains obscure.

An interesting eye anomaly that is seen in some of the freshwater dolphins, particularly the Ganges River Dolphin, is microophthalmia and apparent blindness. This dolphin lives in the Ganges, Indus, and Brahmaputra Rivers of India, where they are known as susu and blind dolphin.[6] They feed in the muddy bottoms of the rivers, by raking the bottom with their rostrum, subsisting on shrimp and fish they are able to catch this way.

The eyes of these animals appear to have undergone a regression during their adaptation, and there is a very minimal slit-like opening between the eyelids, onto the surface of a microophthalmic eyeball (Figure 12). Histologic examination of the eye reveals the presence of a thick cornea like structure, with iris and ciliary body structures, and a retina and optic nerve (Figure 13). The eye apparently is nonfunctional. Other species of dolphins which have similar microophthalmic eye development are other freshwater dolphins such as the Amazon dolphin and the Chinese lake dolphin. Another of the freshwater dolphins, the LaPlata River dolphin, also has lost much of its' skin pigmentation. They are usually pale and some nearly white in appearance. These freshwater dolphins are believed to have adapted to their environment from marine cetacean ancestors, and some live remote distances from the ocean, such as the Chinese lake dolphin which lives in a lake at the headwaters of the Yangtze River of China, or the Amazon River, a thousand miles or more from the ocean.

IV. RESPIRATORY SYSTEM

Noninfectious diseases of the respiratory tract appear to be of uncommon occurrence in marine mammals. Parasitic, bacterial, and mycotic infections have all been reviewed, as well as neoplasms of the lung. Degenerative diseases of the marine mammal airways, such as emphysema, chronic bronchitis, obstructive pulmonary disease, and pulmonary fibrosis appear not to have been observed in these animals, but then they are not subjected to the airborne environmental contaminants which are the primary cause, along with cigarette smoking, of these diseases in terrestrial mammals including man.

Cetaceans, when stranded on beaches, or removed from their aquatic environment, no longer have the cooling effect of the water and are unable to maintain their normal body temperature. They rapidly become overheated and develop heat prostration and die. These animals should be kept shaded and wet down when transporting them, or preferably be in a tank of water. Acute death from heat prostration is not accompanied by many specific lesions, since death is by vascular collapse. On examining the lung, they will be heavy and wet, with watery fluid oozing from the cut surface of the lung parenchyma. The trachea and bronchi will be found to contain considerable frothy watery fluid (Figure 14) which freely runs out on opening these airways. Histologic examination of the lung tissue will reveal vascular congestion, and flooding of the bronchioles and alveolar spaces by edema fluid (Figure 15).

In the California sea lions we studied, there was one young adult animal which had a pulmonary disease which had histologic manifestation suggestive of an allergic or hypersensitivity basis. There was a chronic mucoid bronchitis, with a collection of mucous mixed with inflammatory cells in the lumen of all of the bronchi and bronchioles (Figure 16). There was a hyperplasia of the bronchial glands and a marked goblet cell hyperplasia in the bronchial mucosa (Figure 17), as well as a goblet cell metaplasia in the bronchial glands.

The inflammatory cell component was quite interesting and consisted of a few lymphocytes and plasma cells, but the majority of the inflammatory cells were eosinophils. There was a marked infiltration of mature eosinophils into and around the wall of the pulmonary vasculature (Figure 18), as well as around the bronchi and bronchioles. There were many eosinophils mixed in the mucoid exudate within the lumen of the bronchi (see Figure 17), and there was a massive accumulation of eosinophils in some of the alveolar spaces (Figure 19). The capillaries in the alveolar walls were very dilated. The histologic features described have all of the hallmarks of chronic allergic pneumonitis such as is seen in asthma or some of the idiopathic eosinophilic pulmonary diseases of other mammals.

V. DIGESTIVE SYSTEM

The liver was the site of several interesting pathologic conditions worthy of mention. We encountered two cases of cholelithiasis, one a California sea lion and the other an elephant seal. The sea lion had numerous, almost black colored, intrahepatic bile duct calculi, causing occlusion and dilation of those affected ducts (Figure 20). There was glandular hyperplasia, some fibrosis and chronic inflammation in the wall of the obstructed ducts (Figure 21). Chemical analysis of the calculi revealed a mixture of cholesterol and bile salts. The other calculus was in the extrahepatic common bile duct near the entrance into the duodenum (Figure 22). This was a large solitary stone with a quite roughened surface, dark in color on the surface, and it was quite hard. When cut with a saw, it was found to be made up of a series of concentric lamellae of variable thickness. There was a marked hyperplasia of the glands of the lining of the bile duct system, and a mild to moderate chronic inflammation of the common bile duct, the hepatic duct, and the intrahepatic bile ducts.

One sea lion in our series was found to have a markedly diffuse fatty liver, with numerous large blood filled peliosis hepatis type cavities throughout (Figure 23). Whether these resulted from areas of hepatic necrosis and collapse, or from primary vascular dilation is not known.

Fatty liver was a frequent observation in our study of marine mammals. It usually took the form of a diffuse fatty vacuolation of hepatocytes, affecting the centrilobular and midzonal areas of most lobules (Figure 24), with less at the portal areas. The fatty vacuoles varied from multiple small cytoplasmic ones to large coalescing vacuoles filling the cytoplasm. This fatty change in the hepatocytes may have been the result of inanition associated with intercurrent disease and inappetance, or it might have been a primary metabolic disorder. The high protein diet of these animals may make them prone to fatty change in the hepatocytes, such as occurs in poultry. The fatty change of these livers made them have a pale appearance, and they were sometimes swollen and friable if quite fatty.

Occasional small hyperplastic nodules could be found in the livers of sea lions, and in these foci there would be a different, usually more eosinophilic tinctorial characteristic, and the hepatocytes would be smaller, more closely packed, with a loss of normal trabecular pattern (Figure 25). These were similar to the nodules seen in the liver of rodents following exposure to various chemicals, and may have had a similar pathogenesis in the sea lions.

Pancreatic disease has been described in both cetaceans and pinnipeds. This can take the form of acute pancreatitis,[6] possibly due to ascending infection from the duodenum, although this is speculative, or chronic fibrosing pancreatic disease observed in older cetaceans.[7]

We have observed several cases of acute pancreatitis in sea lions which stranded and died. They had lesions similar to those seen with acute pancreatitis in the dog or other mammal, including focal acute pancreatic necrosis, with leakage of pancreatic enzymes into the peripancreatic fat, causing fat necrosis and saponification, with an acute inflammatory cell infiltrate (Figure 26).

Severe pancreatic atrophy was seen in another sea lion, where the pancreatic tissue was reduced to a thin ribbon of tissue running through the fatty tissue adjacent to the duodenum. Ductal tissue was still recognizable in the tissue section (Figure 27), but there was a negligible amount of zymogen containing acinar cells remaining. This was in an older animal and was therefore interpreted as pancreatic atrophy rather than hypoplastic development of the tissue.

VI. GENITOURINARY SYSTEM

Cystic kidneys have been observed in dolphins and sea lions, in the absence of any evidence of obstructive lesion, and were considered to be developmental.

The kidney of one dolphin contained numerous varying sized cystic cavities up to 1 cm in diameter, scattered randomly throughout the kidney (Figure 28). The larger cystic spaces were often surrounded by a thick whitish wall of dense fibrous connective tissue, and on cutting through the cysts they were filled with clear watery fluid and occasionally with a thin gelatinous material. On histologic examination of tissue sections, the cystic spaces were enclosed in a fibrous capsule, and were encroaching on the adjacent renal cortical tissue (Figure 29).

One 5-year-old male California sea lion was found to have a solitary cyst about 15 cm in diameter located at the anterior pole of the right kidney. It was fluid filled and contained fibrinous and thin gelatinous material (Figure 30). Another sea lion had multiple randomly scattered small cysts located throughout the renal cortices, much like what was seen in the dolphin except that these cysts were not surrounded by a zone of fibrous connective tissue (Figure 31). It appeared there may have been other developmental anomalies in the kidneys of this young adult animal, as many of the glomeruli were seen to be poorly developed or rudimentary (Figure 32), and many of the tubules had not completely developed.

Mineralization within the kidney was occasionally seen, and sometimes was detected as discreet mineral deposits within the lumen of the tubules and tubular epithelium in the renal papillae, such as is seen in the kidney of a beaked whale in Figure 33. These types of intratubular papillary mineral deposits are also seen in dogs, and may be caused by an imbalance in the calcium to phosphorus ratio in the diet, with an excess phosphorus intake.

Multiple renal calculi have been observed in a harbor seal, partially or completely filling many of the renal calyces (Figures 34 and 35), and causing some dilation of tubules and probably some functional impairment of the kidney.

Idiopathic nephritis, other than that associated with leptospirosis or other identifiable infectious agents, was found occasionally in our marine mammal survey. This sometimes took the form of subacute to chronic interstitial nephritis with an interstitial and periglomerular infiltrate of lymphocytes and plasma cells (Figure 36), usually with minimal or no fibrosis. This could represent a remote, and chronic active infectious nephritis, although this could not be proven.

Primary glomerular disease is uncommon in our survey, with only two animals showing evidence of diffuse glomerular damage. A beaked whale had a moderate membranoproliferative glomerulopathy, with uniform thickening of Bowman's capsule, and an increased cellularity of the glomerular tufts (Figure 37). A California sea lion, which died after stranding, was found to have a much more severe glomerular disease, with severe membranous thickening, hypertrophy of parietal epithelial cells, atrophy of glomerular tufts, and total glomerulosclerosis of some affected glomeruli (Figure 38). The etiology and pathogenesis are obscure, but it would be unlikely if these animals escaped the comparable kidney disease problems of terrestrial mammals, such as immune complex glomerulonephritis, which could result in the severe damage such as seen in this case.

Infections of the uterus were described in the chapter on bacterial diseases and will not be mentioned here. Chronic granulomatous orchitis was observed and described in an Amazon River dolphin, but the etiology was not determined.[1] The affected testicle was swollen, turgid, and hyperemic (Figure 39), and on histologic examination it was found that there was a multifocal to diffuse interstitial mixed granulomatous inflammatory infiltrate, with some necrosis in the center of the granulomas (Figure 40).

VII. ENDOCRINE SYSTEM

Tumors affecting endocrine glands have been seen rather frequently and are listed in the chapter on neoplastic diseases, but other diseases of the endocrine system have been infrequently reported. Cowan[8] described colloid goiter in four pilot whales and granulomatous thyroiditis in two pilot whales. As Ridgway and Patton have reported,[3] the thyroid glands of cetaceans are considerably larger in relation to their overall weight, as compared to terrestrial mammals.

In our study, we had one northern right whale dolphin which had somewhat enlarged and mushy thyroid glands. On cut section it was found that the softness was due to diffuse cystic degeneration of the thyroid glands. On histologic section, numerous varying sized cystic structures were evident (Figure 41). Some were lined by flattened thyroid acinar epithelial cells, and most cysts were filled with inflammatory and cellular debris. On closer examination, it was determined that there was a diffuse chronic lymphocytic interstitial thyroiditis (Figure 42), and it is presumed that a long-standing inflammatory, and possibly immune mediated disease of the thyroid glands resulted in the severe cystic degeneration observed.

Another thyroid gland lesion which was observed incidentally in a northern right whale dolphin was a nodular hyperplastic proliferation of epithelial cells, which had the organoid appearance characteristic of parafollicular (C - cells) cells (Figure 43). These cells are responsible for elaboration of calcitonin.

A rather frequent finding in our survey of stranded animals was a moderate to severe degeneration of the adrenal cortex. This was always a bilateral lesion, and usually consisted of a vacuolar degeneration of glomerular and reticular zones of the adrenal cortices (Figure 44). This type of degeneration is commonly considered to be a nonspecific lesion associated with stress, disease states, or trauma.

VIII. HEMATOPOIETIC SYSTEM

Other than neoplastic disease of the hematopoietic tissue described in the chapter on neoplasms, several entities affecting this tissue will be mentioned.

One older common dolphin in our series was found to have, as an apparent incidental lesion, amyloidosis affecting only the lymphoid germinal centers of the spleen. Most all of the follicles on the sections examined were affected (Figure 45), with fairly discreet round to slightly oval deposits of amorphous pink proteinaceous material. This material had the appearance of amyloid (Figure 46), and was positive on the Congo Red stain and by birefringence as amyloid. None of the other tissues examined showed evidence of amyloidosis. It has not been possible to examine this further to determine if this is of immunoglobulin origin or contains specific amyloid protein. It seems likely that this lymphoid associated amyloid deposition was caused by a chronic or prolonged immune response. There was considerable plasma cell differentiation in some areas of the spleen and lymph nodes (see Figure 45, inset).

Other splenic lesions encountered in several older California sea lions was a moderate diffuse myeloid metaplasia in the red pulp (see Figure 46). There appeared to be a trilineage of differentiation, with megakaryocytes, erythrocytes, and granulocytes, with their respective undifferentiated precursor cell types. Since the marrow of these animals does undergo atrophy with increased trabecular bone deposition in the marrow cavity, this may simply be a normal physiologic response. There was no evidence of myeloid metaplasia in other tissues.

Occasional subcapsular lesions were seen grossly in the spleen, and these were yellowish white discolored plaque-like areas involving the capsule. Histologically, these were areas of prior hemorrhage that were undergoing resolution and scarring, with residual blood pigments causing the discoloration (Figure 47). These were the same as the siderotic plaques, or so-called Gandy-Gamna bodies that are frequently observed in the spleens of old dogs. If numerous or extensive, they may be indicative of a bleeding disorder.

IX. CARDIOVASCULAR SYSTEM

Degenerative vascular disease has been described in cetacea, and varied from small fibrous intimal plaques in the coronary arteries of a porpoise,[1] to larger and multiple fibrous plaques in the aorta which had traces of sudanophilic material deep in the plaques.[9] Larger, and frankly atherosclerotic coronary artery lesions were also described in an aged killer whale, and possible early atherosclerotic aortic lesions in a baleen whale.[9]

In the group of pinnipeds we have examined, we did not encounter evidence of degenerative vascular disease affecting the coronary arteries or aorta. However, in several older dolphins we found occasional small aortic lesions similar to what has been described. These lesions consisted of fibrous intimal proliferations forming intimal plaques (Figure 48). There was disruption and reduplication of the internal elastic membrane, and on close examination of the plaques there were focal accumulations of large ''foam cells'', either macrophages or attenuated smooth muscle cells, which were filled with lipid material (Figure 49). These were incidental lesions, apparently causing no obstruction, and simply indicate that these animals may develop mild degenerative vascular lesions, as do their terrestrial mammal counterparts such as the dog, cow, and pig.

We had one Dall's porpoise in our collection which had interesting but idiopathic aortic lesions. These consisted of a very severe diffuse acute aortitis, affecting the wall of the aorta toward the intimal surface (Figure 50), as well as the mid-zone of the wall, along the vasa vasorum (Figure 51). The lesions were fairly extensive along the aorta, and consisted of a mixture of inflammatory cells, including neutrophils, lymphocytes, and macrophages. Whether these lesions were part of a specific disease entity, or were simply due to a low grade septicemia, was not determined.

Focal myocardial scarring has been described in the dolphin,[1] where the lesions may have been the end result of small myocardial infarcts. We have seen similar small foci of myocardial fibrosis in several mature California sea lions which stranded and died, where these lesions were probably incidental to the cause of death of the animals. We have also seen other myocardial lesions in other animals, which lead us to believe that some of these focal areas of myocardial fibrosis may be sequelae to myocarditis, and possibly a specific type of myocarditis, perhaps with a viral etiology. An example of this was found in an adult female sea lion, which at necropsy examination was found to have multiple small pale foci randomly scattered in the myocardium. Histologic examination of these areas revealed that there was myocardial fiber degeneration and necrosis, with fibrous connective tissue replacement (Figure 52). In some other, and apparently earlier lesions, there were infiltrates of lymphocytes and monocytes among the degenerating muscle bundles (Figure 53), suggesting a possible infectious etiology. Proven cases of infectious or viral myocarditis have not been described in marine mammals, but most likely do occur.

Similar focal lesions of myocarditis were found in the hearts of several sea lions that were examined during an unusual mass stranding of 60 to 70 or more California sea lions along the Ventura and Santa Barbara County coastal area about 5 years ago. No consistent pathologic lesions were noted in the group of animals examined, and the cause of that mass pinniped stranding remains undetermined, even though extensive studies were carried out by several laboratories.

Cardiac lesions consisting of necrosis of myocardial muscle fibers, has been reported to occur in vitamin E deficiency myopathy of marine mammals, usually in conjunction with diffuse or multifocal skeletal muscle necrosis.[10,11] The myocardial lesions associated with vitamin E deficiency are quite different from the lesions that we have just described in sea lions. The myocarditis that we have seen contained multifocal inflammatory cell infiltrates, predominantly mononuclear inflammatory cells, in conjunction with myocardial fiber necrosis, and then mostly just fibrosis in the older lesions. There was no evidence of fragmentation or calcification of muscle bundles, or proliferation of sarcolemmal nuclei, or hyalin degeneration of sarcoplasm, all of which are lesions more characteristic of vitamin E deficiency myopathy.

Severe necrotizing myopathy, affecting skeletal muscles, has been seen in two stranded sea lions of our series. Death occurred 3 to 7 days after they became stranded, and were transported for treatment and observation. The skeletal muscles over the lumbar region revealed severe acute hemorrhage and necrosis of muscle bundles (Figure 54), without any inflammation or regeneration. No kidney lesions were found in these animals, but these muscle lesions are believed to be consistent with what has been described in other wild mammals as "capture myopathy", or acute exertional rhabdomyolysis. Possible mechanisms for the development of this lesion has been reviewed by Spraker.[10] The same type of lesions are sometimes seen in other wild animals after they are caught or undergo unusual exertion, and the lesion has been seen in man, usually after considerable exercise or exertion, when not in condition, such as may occur in new military recruits.

X. CONGENITAL DEFECTS

In a comprehensive review by Leipold,[12] various congenital defects were described in marine mammals, including such things as ventricular septal defect, transposition of pulmonary artery and aorta, and Hageman factor deficiency in the bottle-nosed dolphin. The latter, Hageman factor deficiency, is felt to be normal in this species. Also reviewed by Leipold were persistent ductus arteriosus, umbilical hernia, and ectrodactyly in harbor seals, penis malformation in an elephant seal, ectrodactyly in a fur seal, deformed forelegs and bilateral twinning of the tusks in walruses, and intestinal disaccharidase deficiency in the California sea lion. This latter enzyme deficiency is normal in the sea lion. Sea lion milk is nearly devoid of any milk sugar such as lactose, and therefore the pups do not have the enzyme to digest such sugars. Their only natural source of carbohydrate consists of small amounts of simple sugars in the fish they eat, and attempts to feed orphan pups an artificial diet containing lactose or sucrose will cause severe diarrhea.

In our experiences, we have observed several animals with congenital defects that we will describe. One young harbor seal, about 4 weeks of age, was stranded and transported for observation and treatment because it was believed to have a respiratory infection. It developed a progressive severe dyspnea and died. At necropsy examination, the pup was found to have right ventricular hypertrophy, and the pulmonary artery conus was markedly dilated. The ascending aorta showed sacculation, and there was a patent persistent ductus arteriosus, about 1/2 cm in diameter, which was the basis for the severe dyspnea and cardiac failure.

Bilateral polycystic kidneys were observed in a common dolphin and a California sea lion, and are described under urinary tract disorders. These were believed to have been congenital anomalies. Unilateral renal hypoplasia was seen in one adult female sea lion, with the right kidney being markedly hypoplastic, and the left kidney showing compensatory hypertrophy. This lesion was incidental, the cause of death being a nonsuppurative encephalitis.

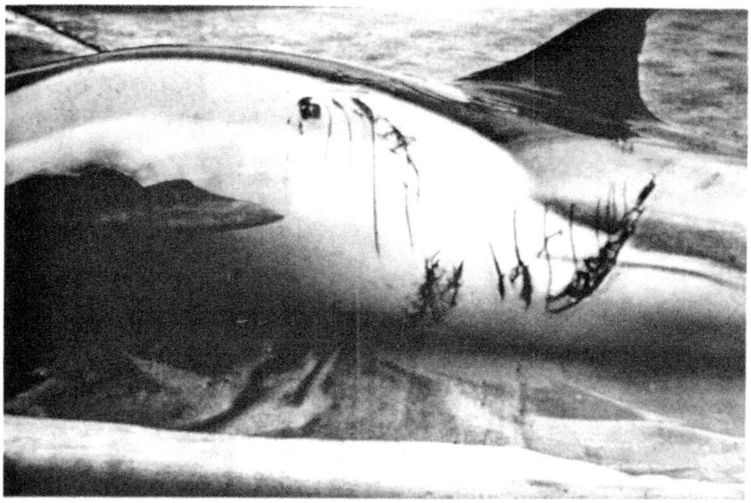

FIGURE 1. Picture of a Pacific white-sided dolphin, showing typical deep wounds characteristic of a shark attack. (Courtesy of Dr. S. Ridgway and the Comparative Pathology Registry, A.F.I.P.).

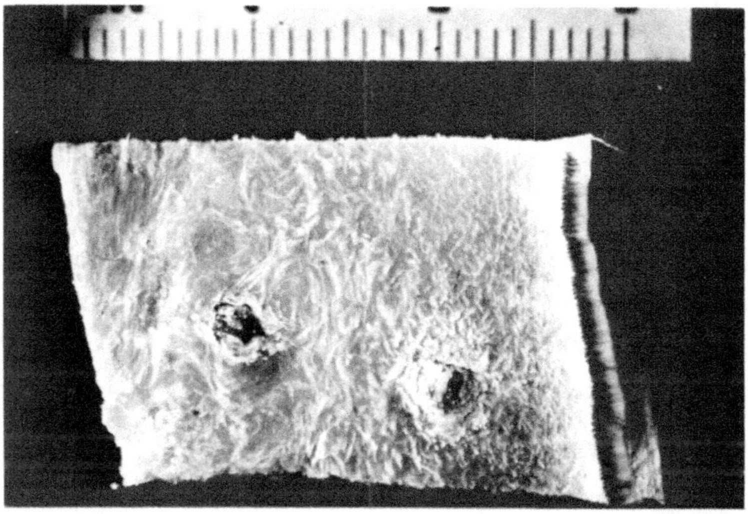

FIGURE 2. Section of dolphin skin, showing deep penetration by two buckshot. (Courtesy of Dr. J. Sweeney and the Registry of Comparative Pathology, A.F.I.P.).

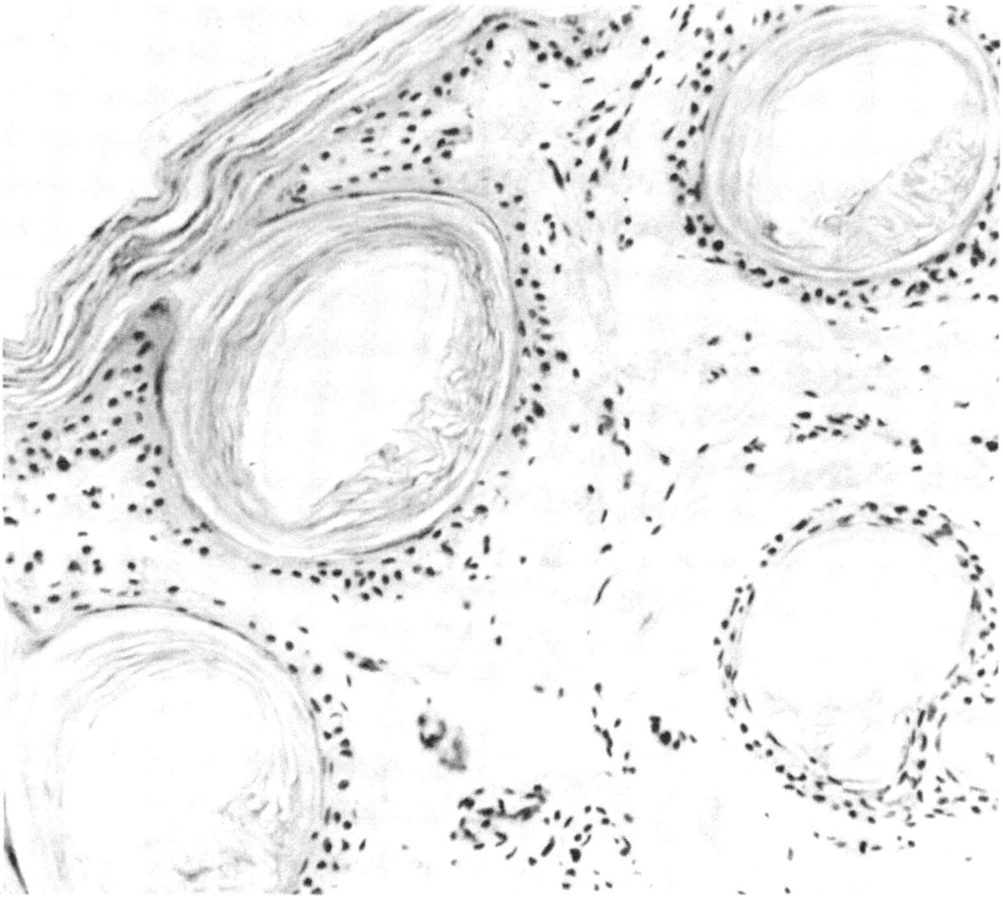

FIGURE 3. Section of sea lion skin, demonstrating hyperkeratosis of the epidermis, with dilation of the hair follicles which are filled with keratin. Possible hypothyroid skin lesions.

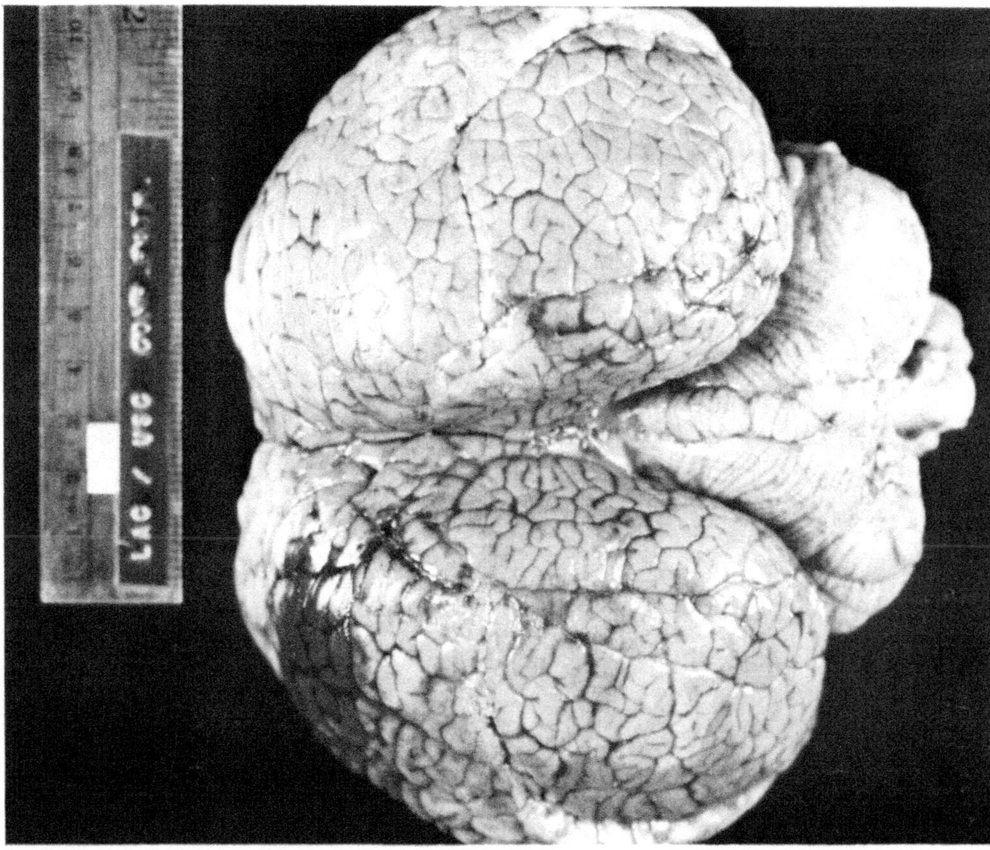

FIGURE 4. Brain from a dolphin showing diffuse subdural hemorrhage over the left anterior cerebral hemisphere, associated with underlying hemorrhagic infarct.

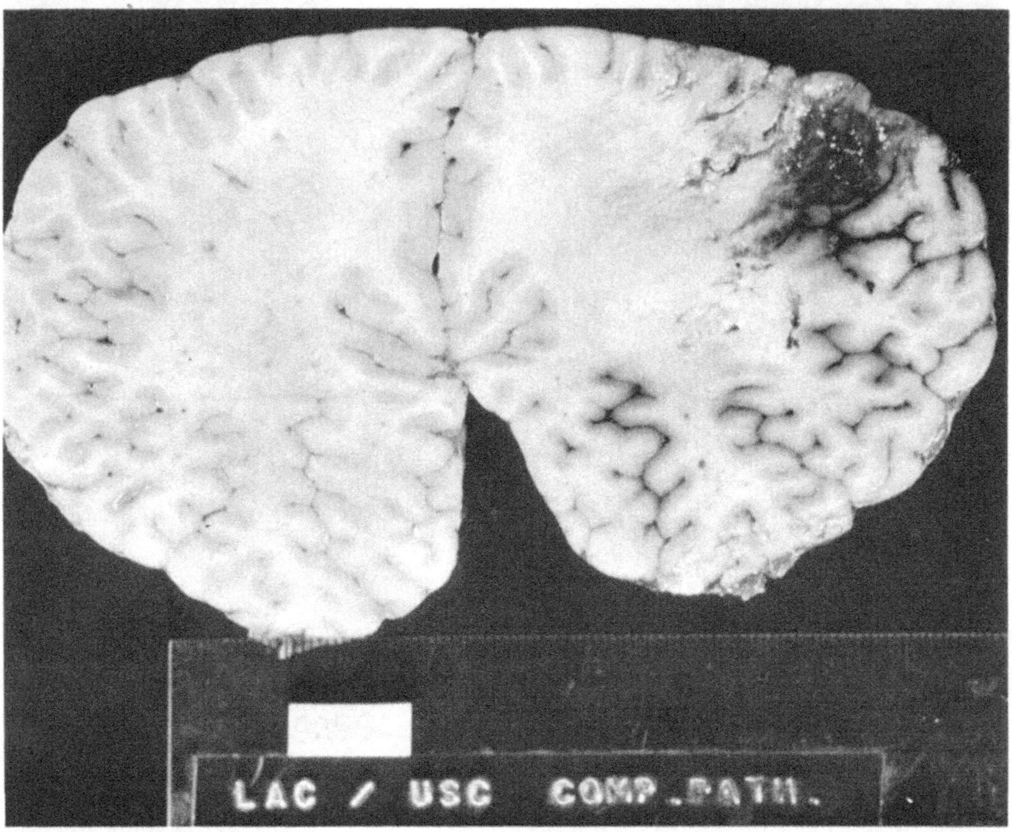

FIGURE 5. Sagittal section through the brain in Figure 4, revealing the extensive hemorrhagic infarct with necrosis.

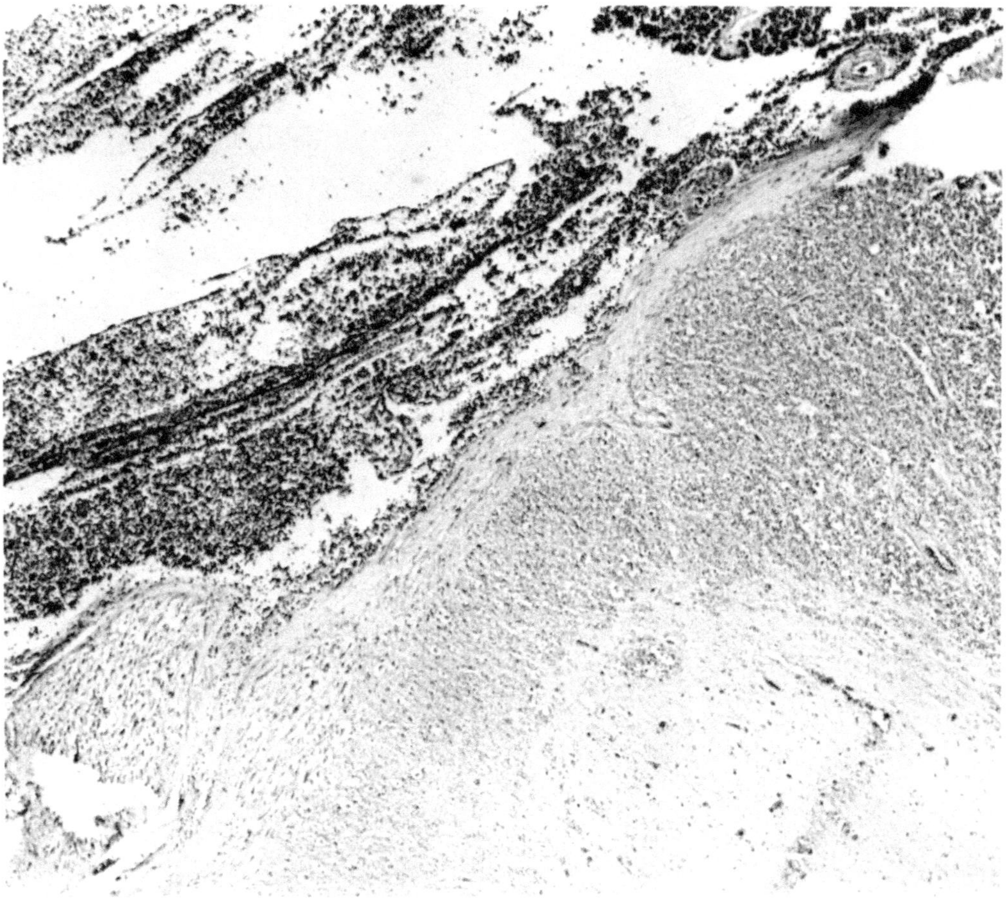

FIGURE 6. Photomicrograph through hemorrhagic area of Figure 4, showing considerable subdural hemorrhage associated with the cerebral infarct.

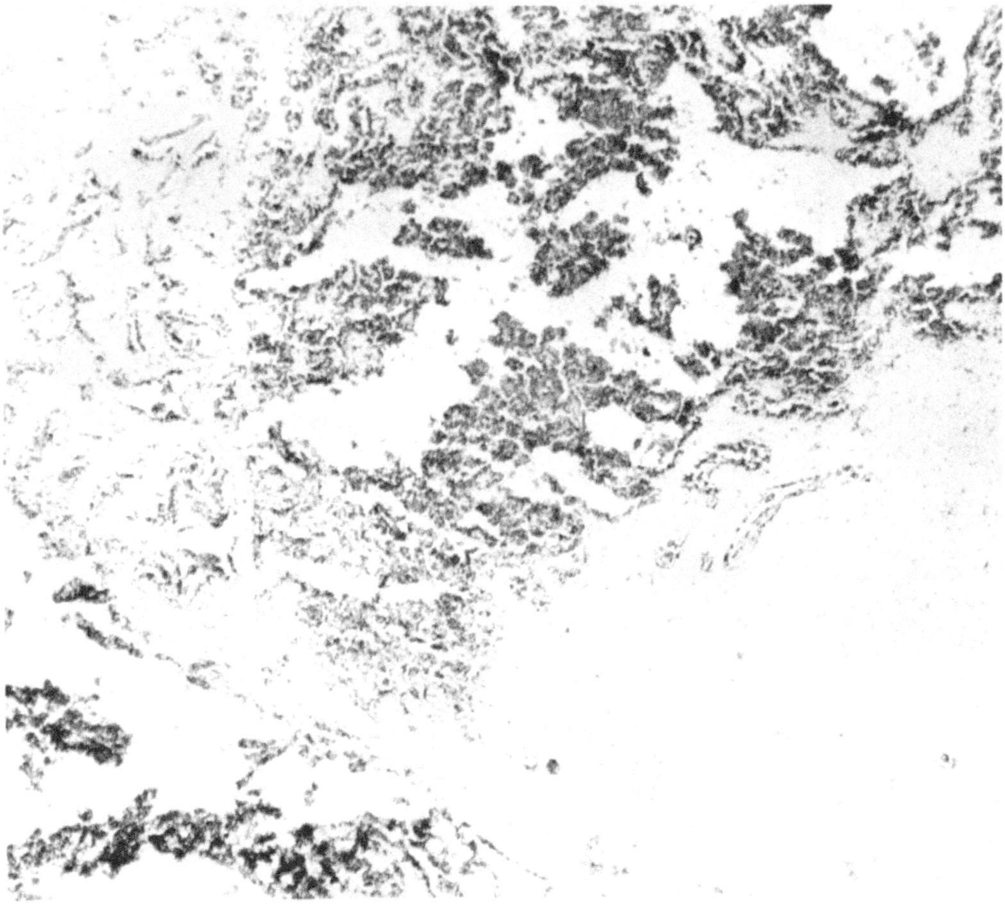

FIGURE 7. Tissue section through area at periphery of hemorrhagic infarct seen in Figure 5, showing extensive hemorrhage and necrosis of brain neuropil.

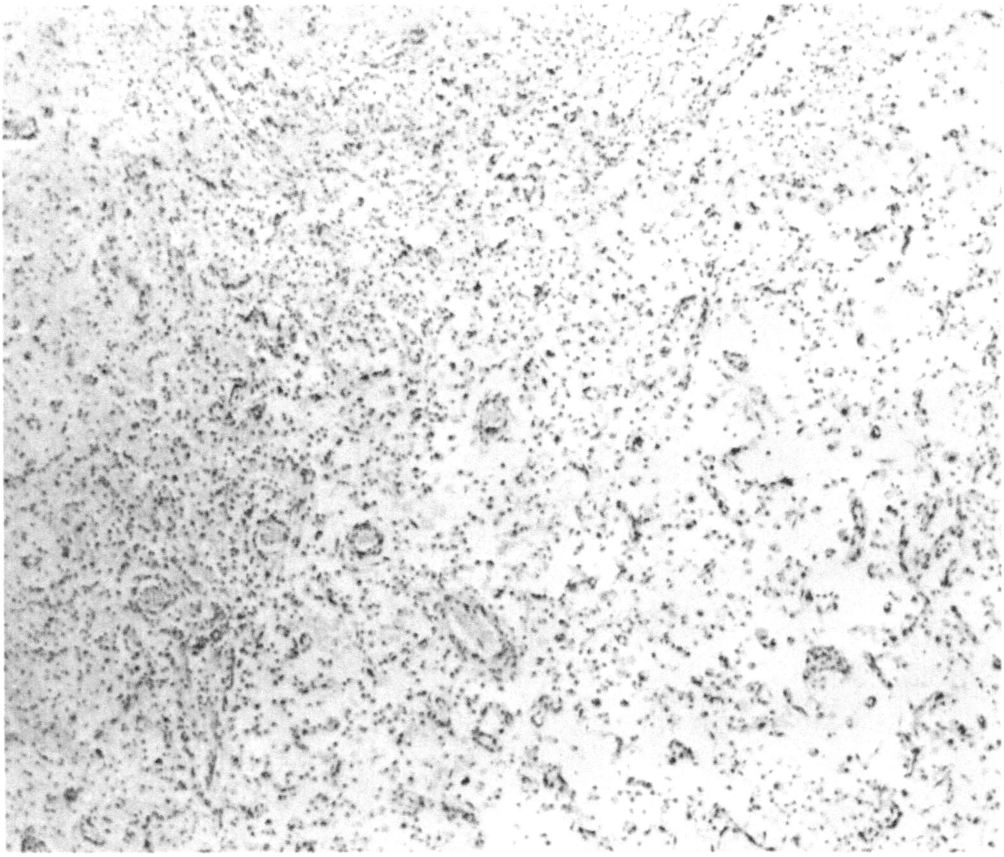

FIGURE 8. Section from the brain of the California sea lion, showing a focal area of liquefactive necrosis, with a central malacic cavity and considerable Gitter cell influx around the edge of the cavity.

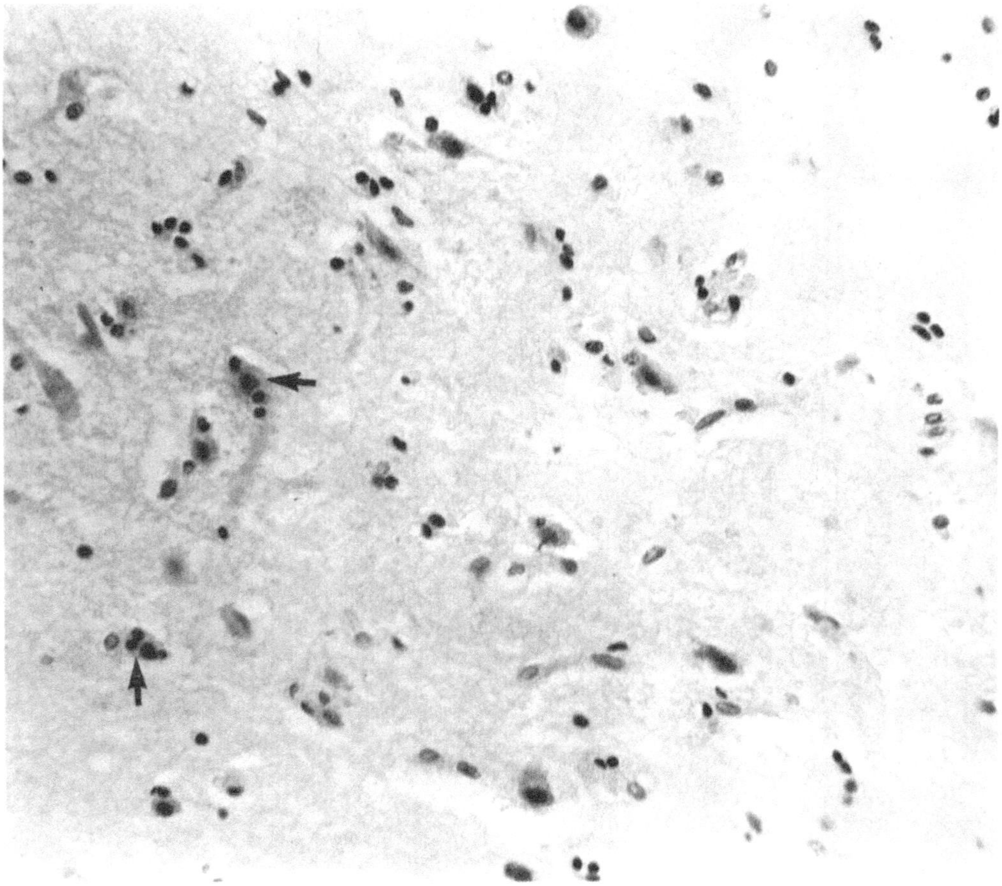

FIGURE 9. Brain section from an elephant seal suspected of having chronic lead poisoning, demonstrating satellitosis and neuronophagia, with pyknosis and degeneration of neuronal nuclei (arrows).

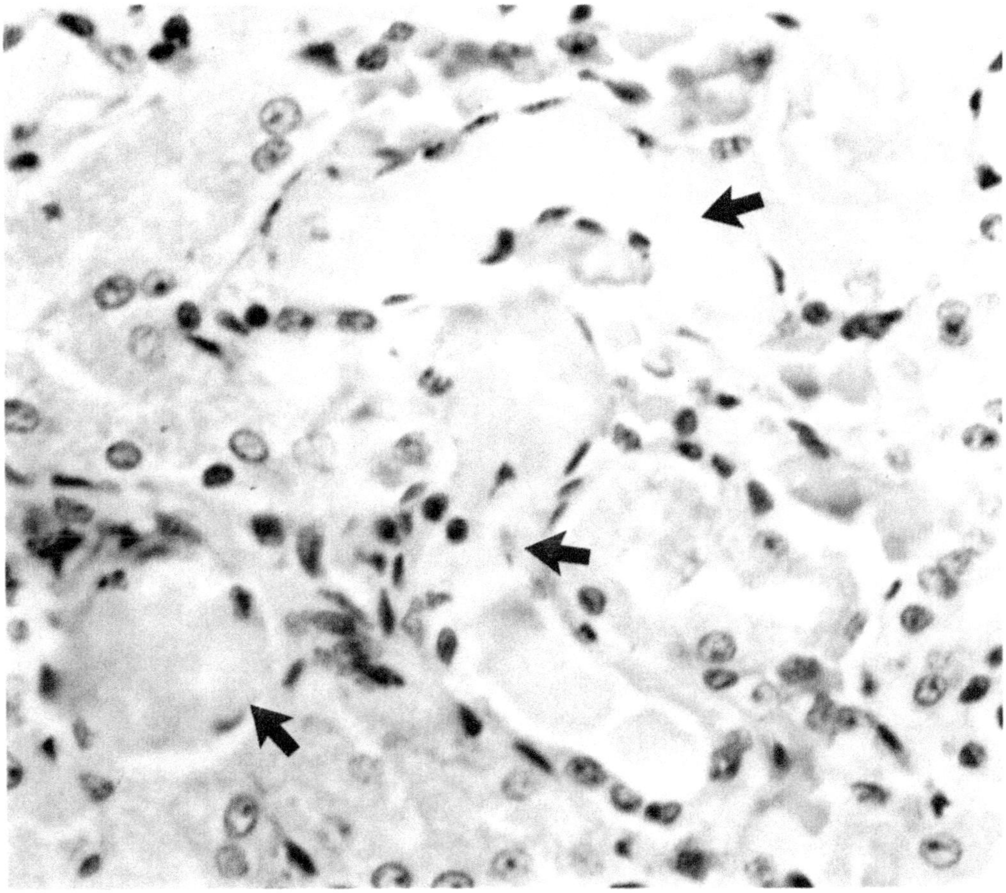

FIGURE 10. Kidney section from the same elephant seal as in Figure 9, demonstrating the marked tubular nephrosis (arrows).

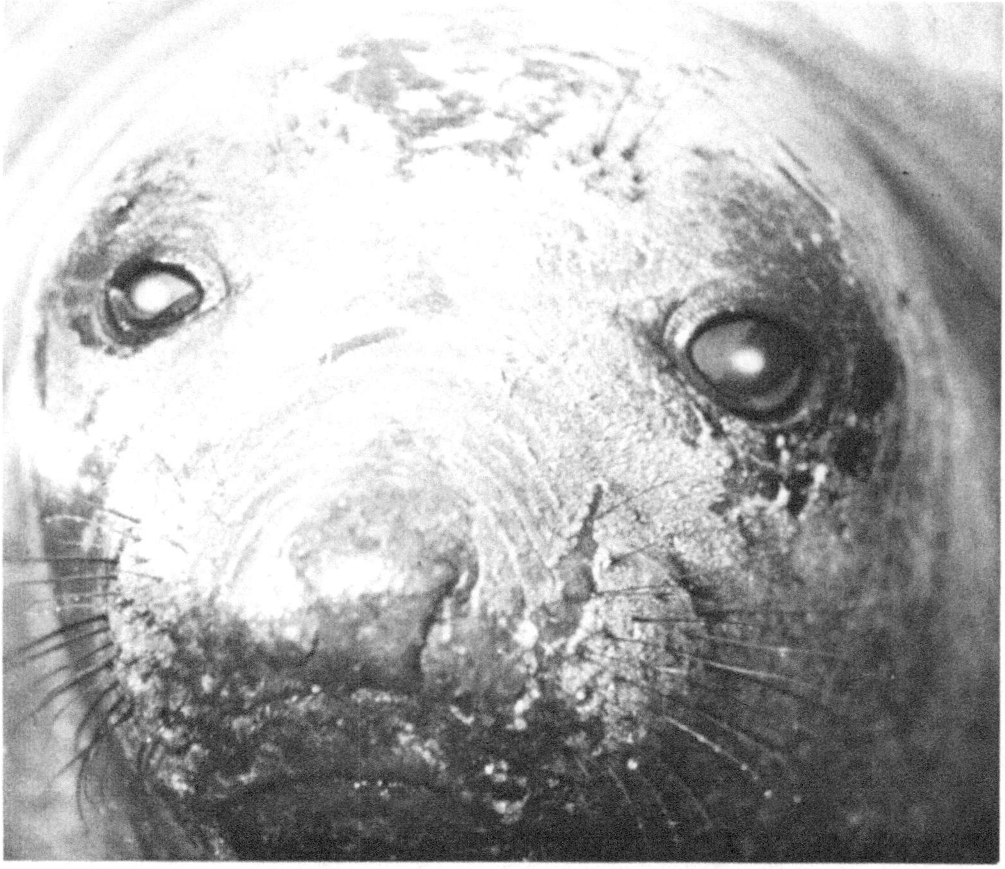

FIGURE 11. Bilateral corneal clouding and opacity in a sea lion pup. The lesions developed after captivity, in a seawater pool.

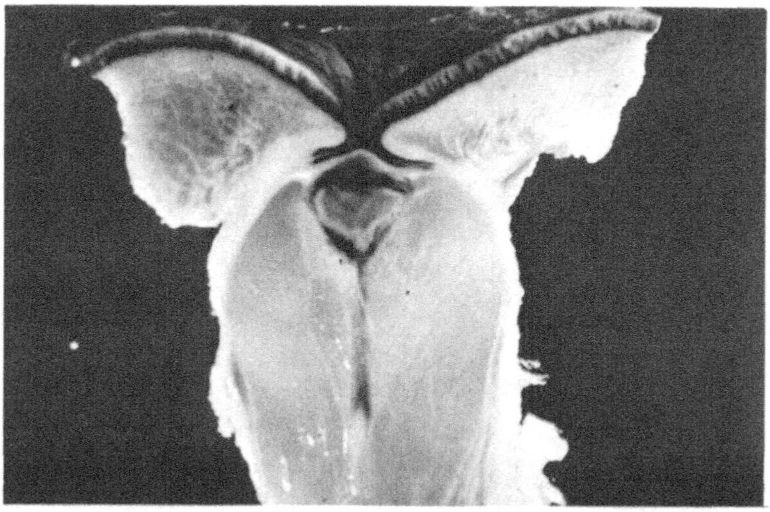

FIGURE 12. Whole section through the eye of a Ganges River dolphin, demonstrating marked microphthalmia. Note the very narrow slit-like opening between eyelids, at top of picture. (Courtesy of Dr. George Migaki and the Comparative Pathology Registry, A.F.I.P).

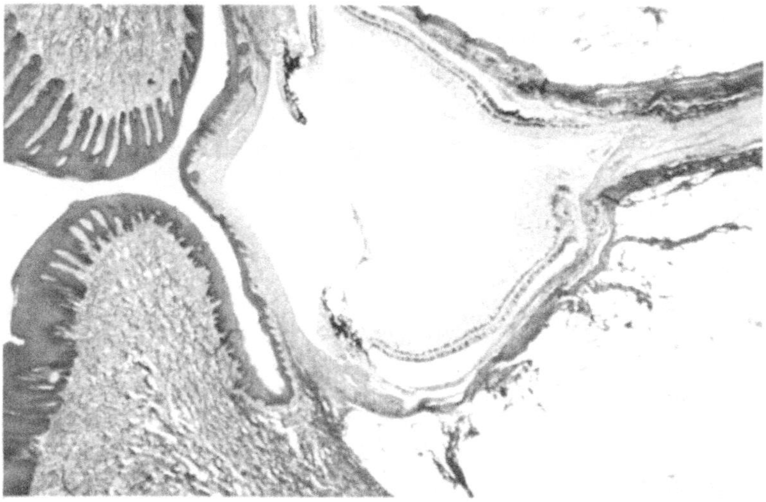

FIGURE 13. Histologic section of the eye shown in Figure 12, showing that the cornea, iris, retina, and vestigal optic nerve are all present in the microphthalmic eyes of the Ganges River dolphin. (Courtesy of Dr. George Migaki and the Comparative Pathology Registry, A.F.I.P.).

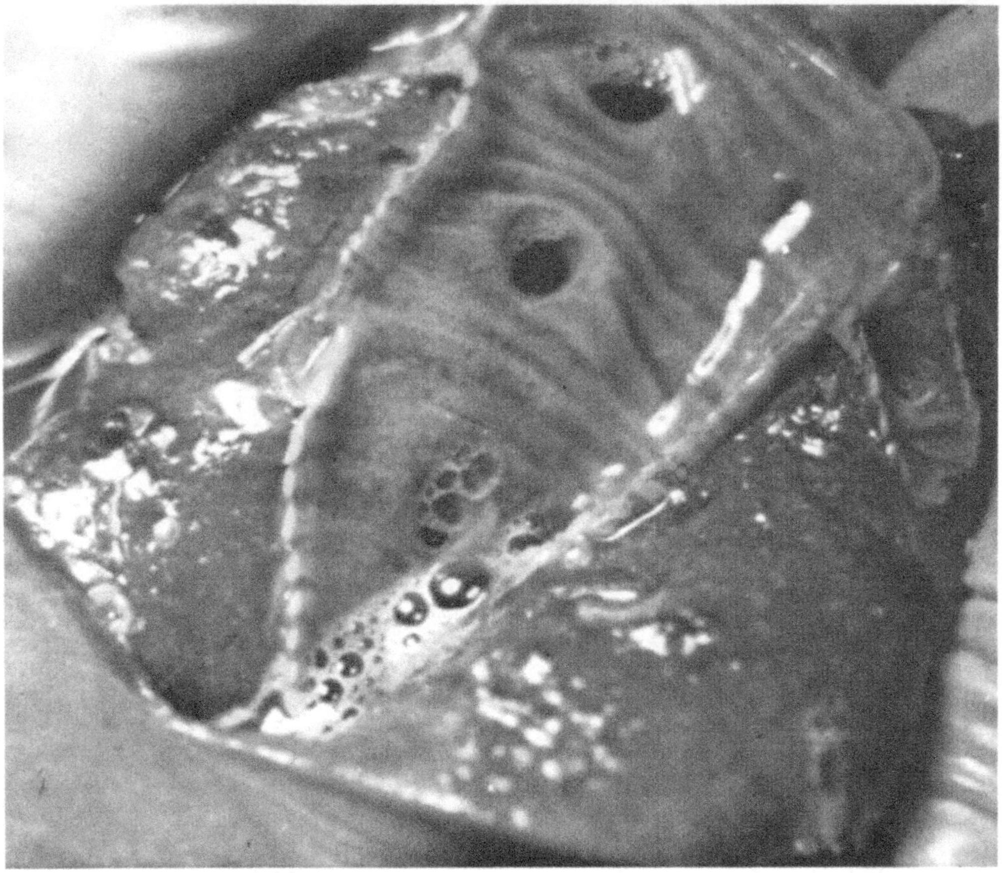

FIGURE 14. Lung from a dolphin that died of heat prostration on the beach. Note the wet, edematous appearance of the cut surface of the lung tissue, and thin frothy fluid in the bronchus.

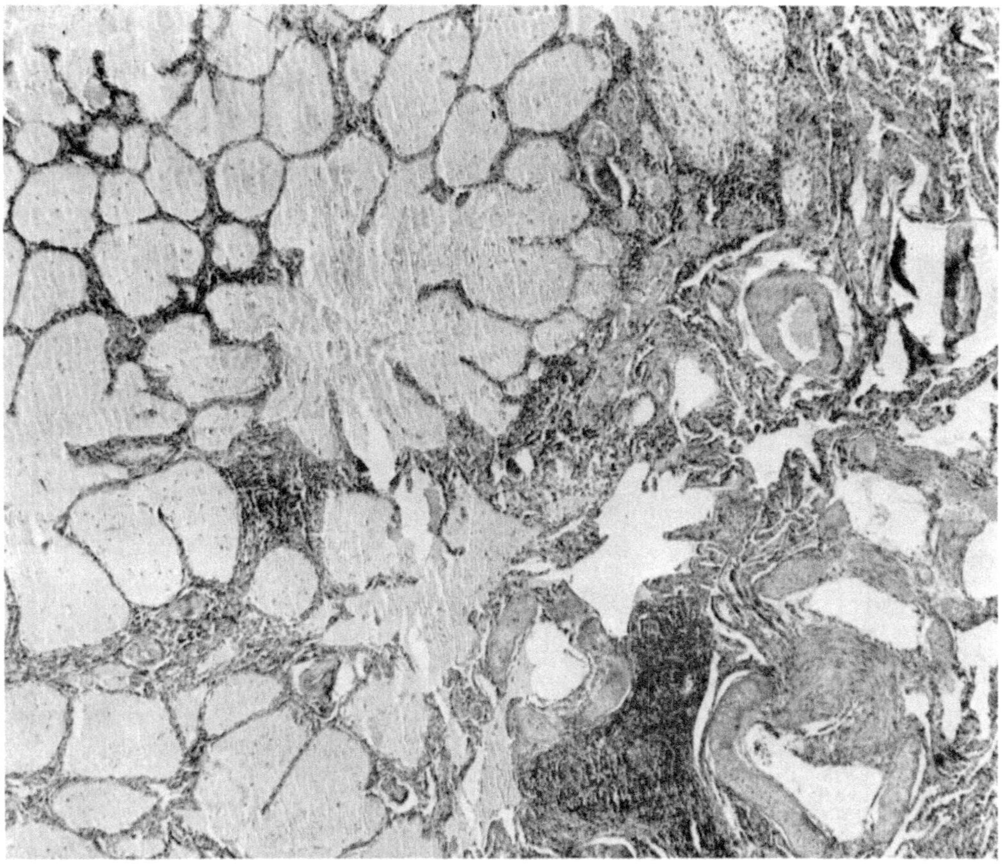

FIGURE 15. Lung section from a dolphin, as in Figure 14, showing the edematous fluid filled bronchioles and alveolar spaces.

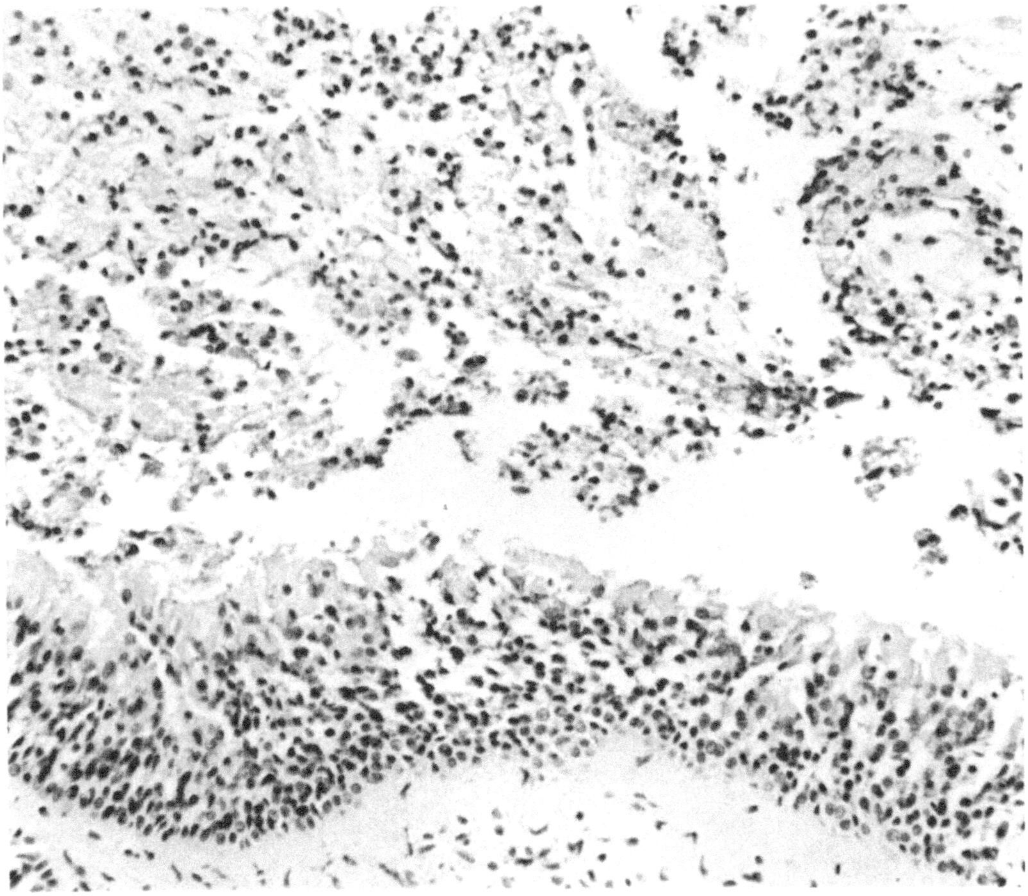

FIGURE 16. Lung section from a sea lion, demonstrating mucoid bronchitis. Lumen of the bronchus is filled with mucous and inflammatory cell debris. The bronchial mucosa appears hyperplastic.

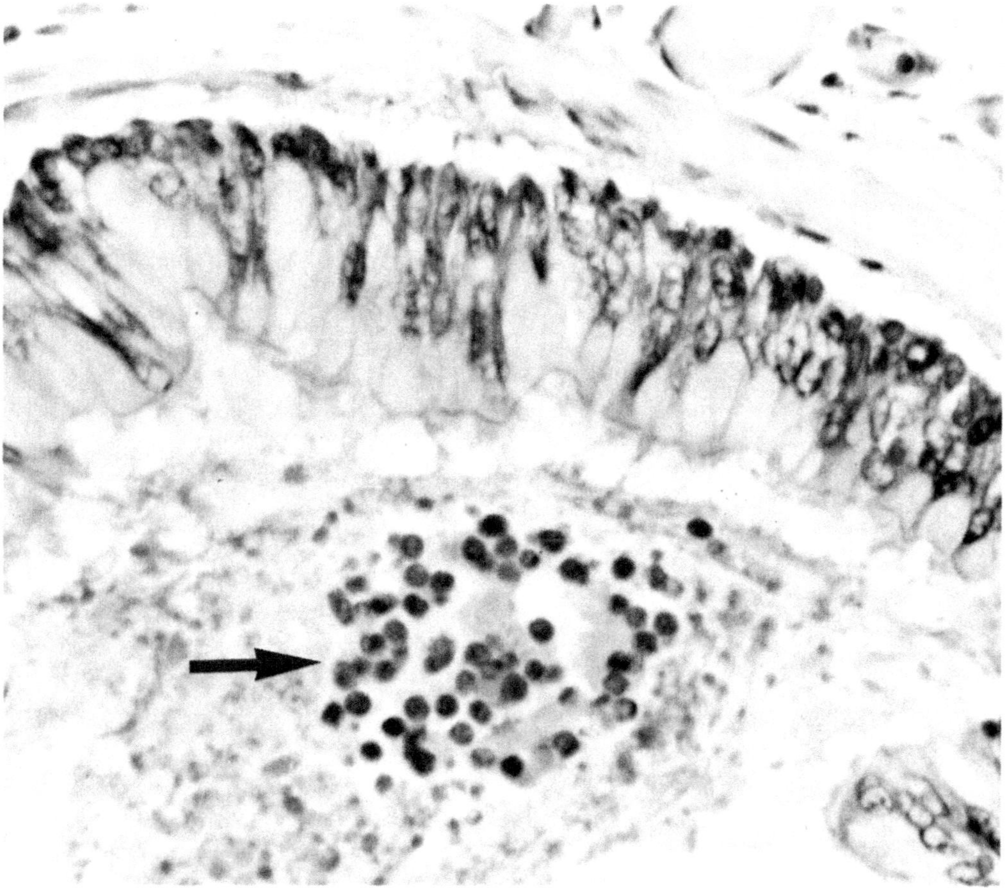

FIGURE 17. Section through the bronchiole of a sea lion with allergic bronchitis. There is considerable goblet cell hyperplasia in the lining bronchial mucosa, and a cluster of eosinophils in a mucous matrix, in the lumen of the bronchiole (arrow).

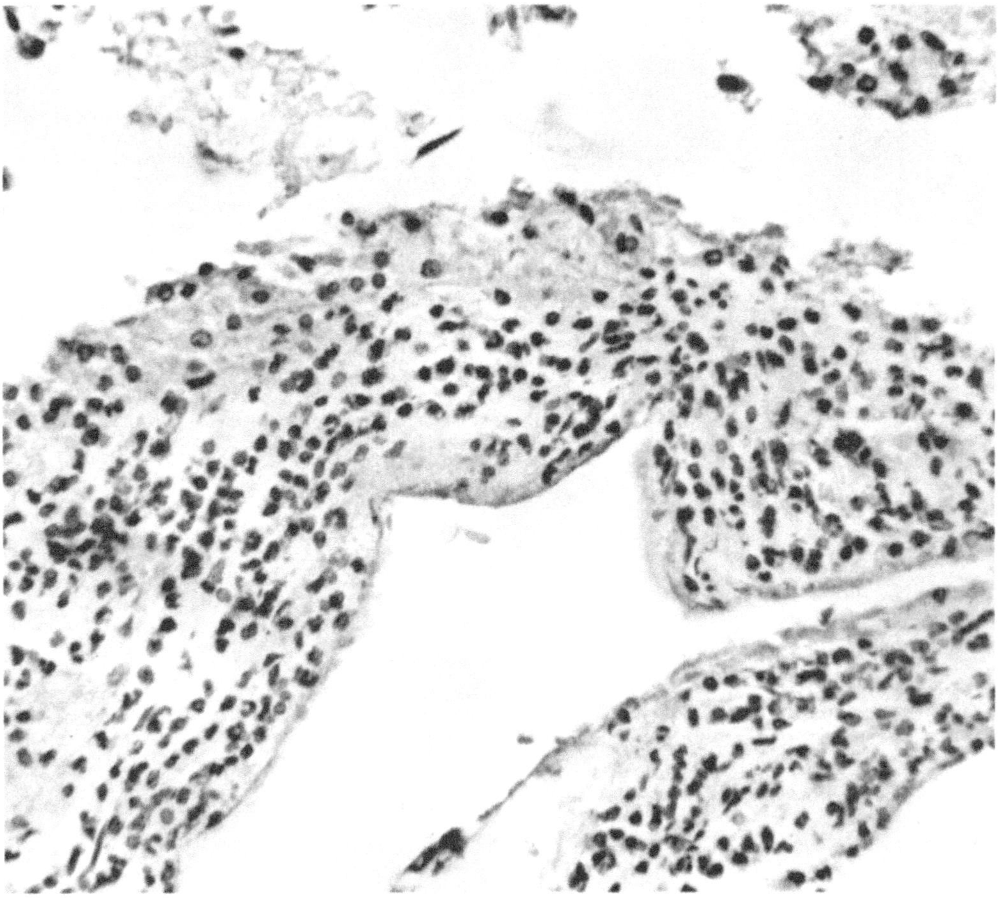

FIGURE 18. This section shows the wall of a pulmonary vessel in the lung of a sea lion with allergic bronchitis, showing a marked infiltrate of eosinophils in the wall of the vessel.

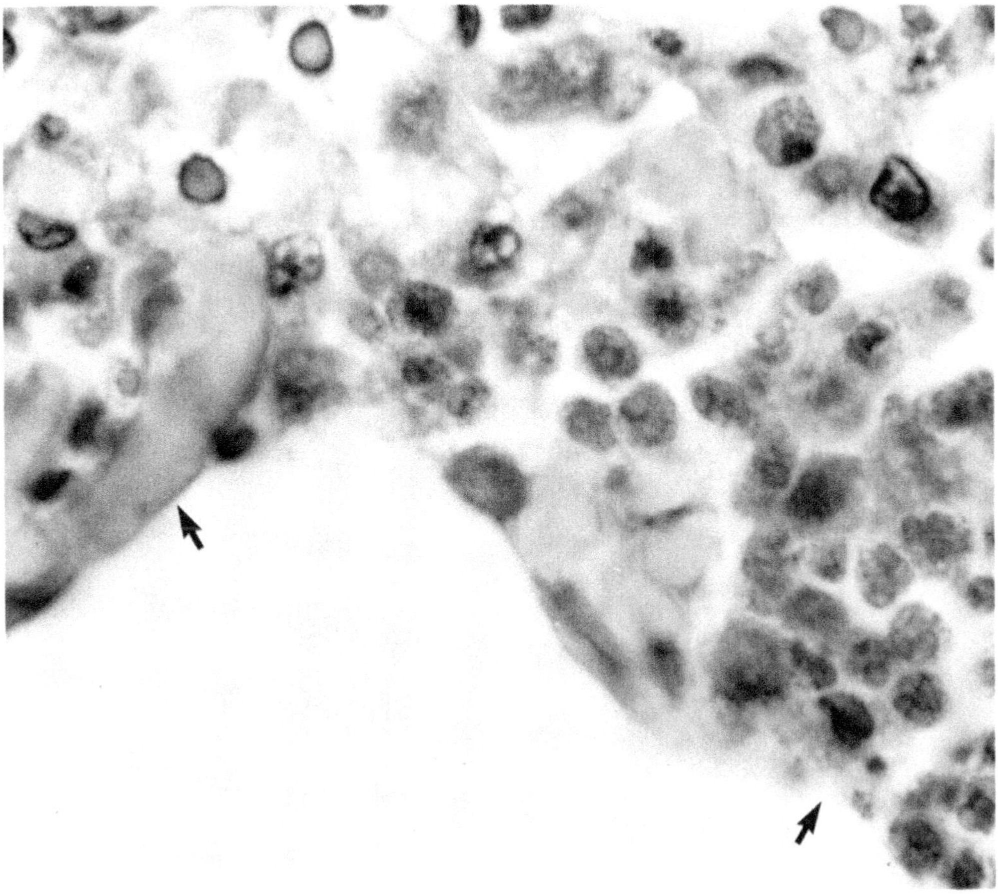

FIGURE 19. Higher magnification of a lung section from a sea lion with allergic pulmonary disease, revealing distended capillaries in the alveolar septa and massive alveolar accumulation of eosinophils (arrows).

FIGURE 20. Cut surface of the liver from a sea lion, demonstrating numerous dark biliary calculi filling the bile ducts, which have markedly thickened fibrotic walls (arrows).

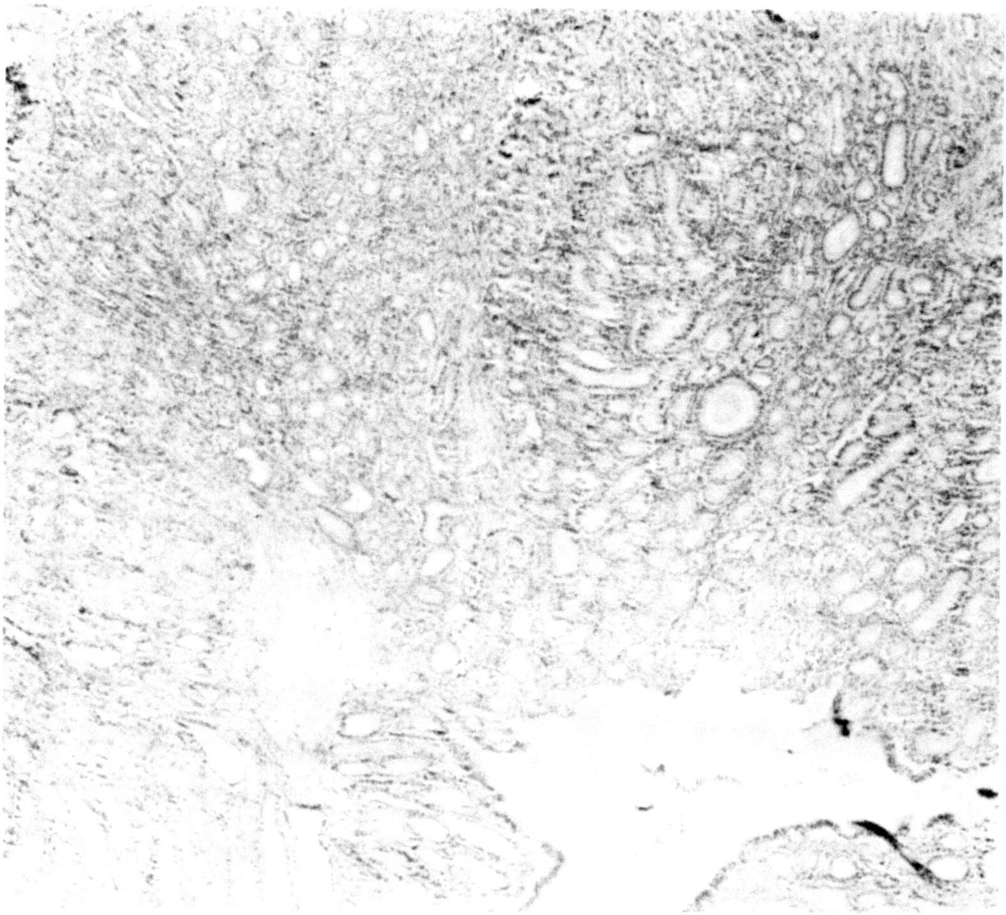

FIGURE 21. Cross-section of a larger intra-hepatic bile duct, revealing the marked hyperplasia of the bile duct lining mucosa, caused by biliary calculi.

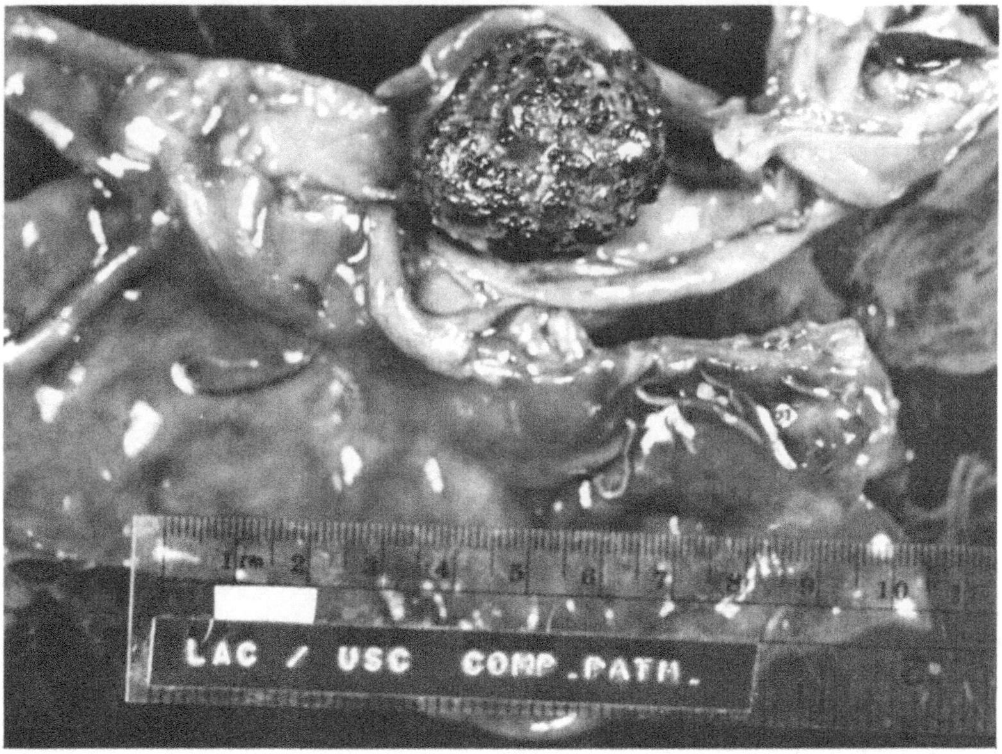

FIGURE 22. Opened common bile duct, adjacent to duodenal opening, from an elephant seal, showing a large irregular surfaced calculus in the lumen of the duct. Obstruction was undoubtedly present.

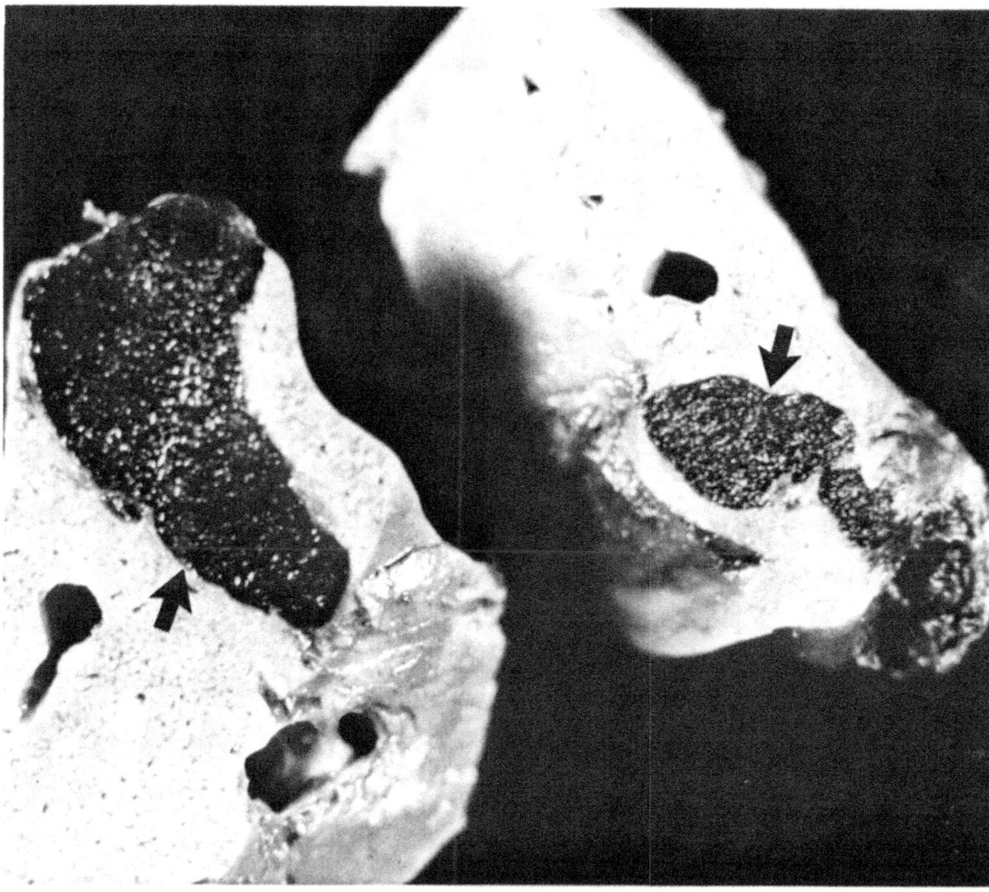

FIGURE 23. This shows cut surfaces of the liver from a sea lion, and it can be seen that the liver is quite fatty, and there are large areas of peliosis hepatis of undetermined etiology (arrows).

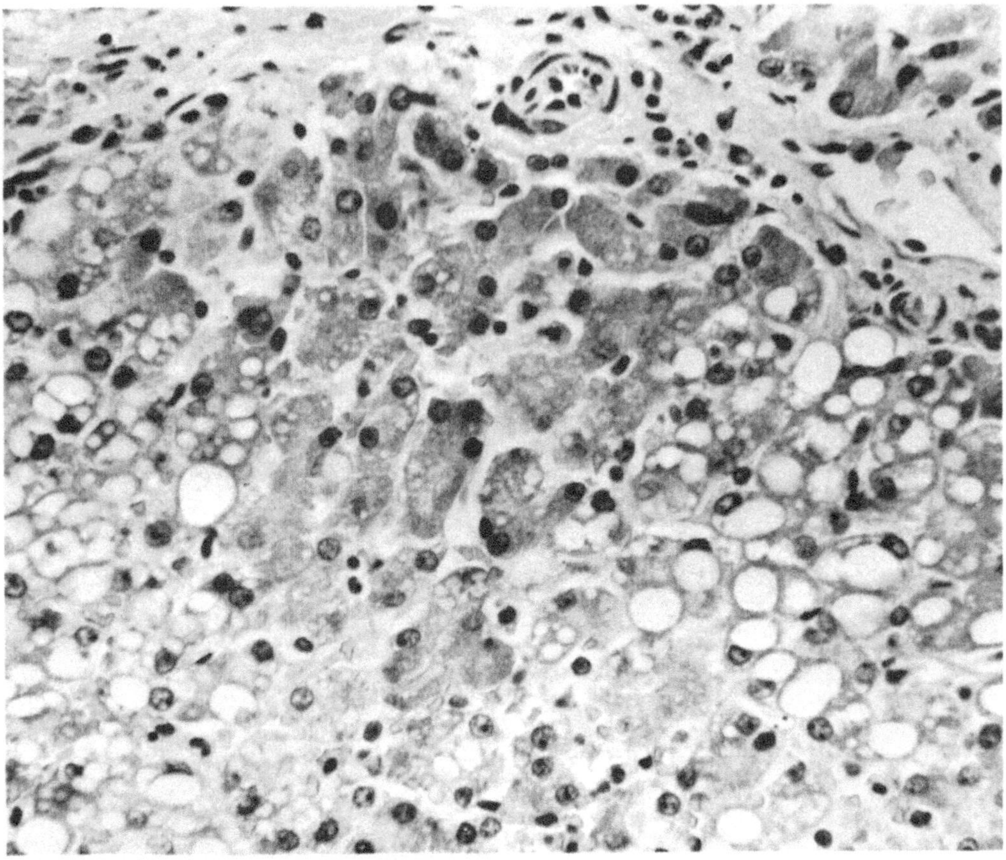

FIGURE 24. Histologic section from the liver of a sea lion, showing a diffuse fatty change, with varying sized lipid vacuoles in the cytoplasm of most hepatocytes.

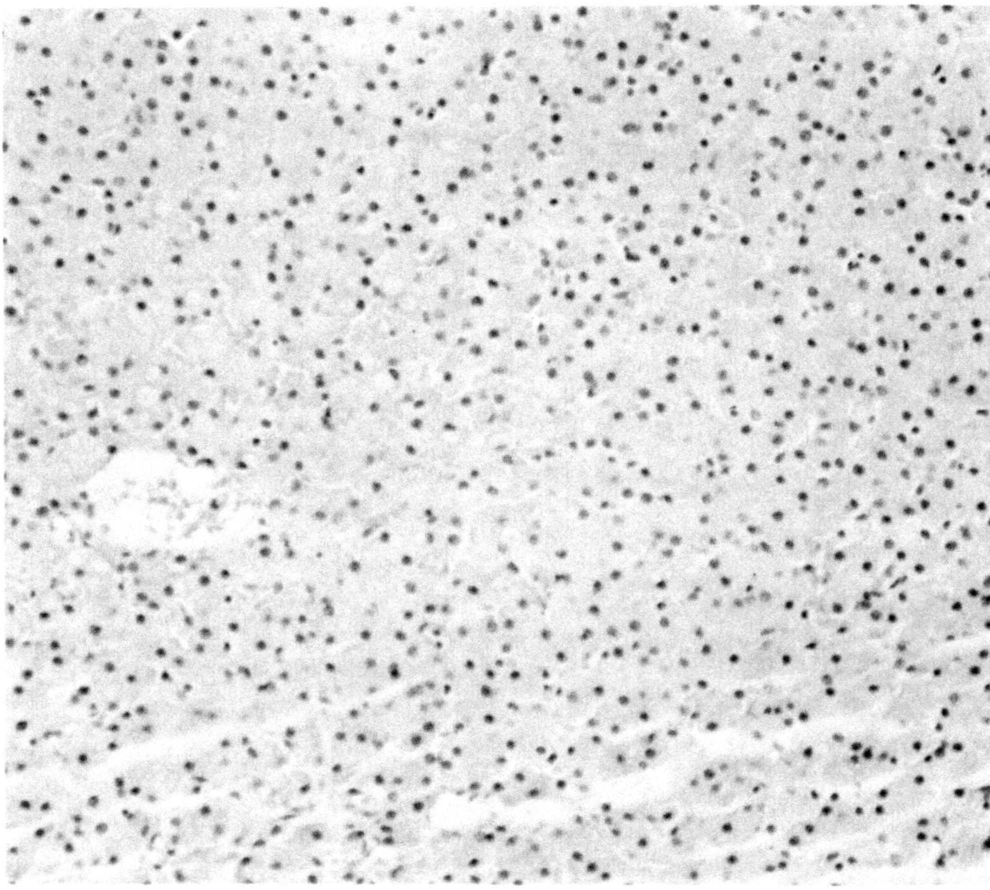

FIGURE 25. This is a liver section from a sea lion, showing a focal hyperplastic nodule, with no well defined margin, and no hepatic cords or sinusoids. Numerous cytoplasmic vacuoles (peroxisomes) can be seen in the hepatocytes in this hyperplastic area.

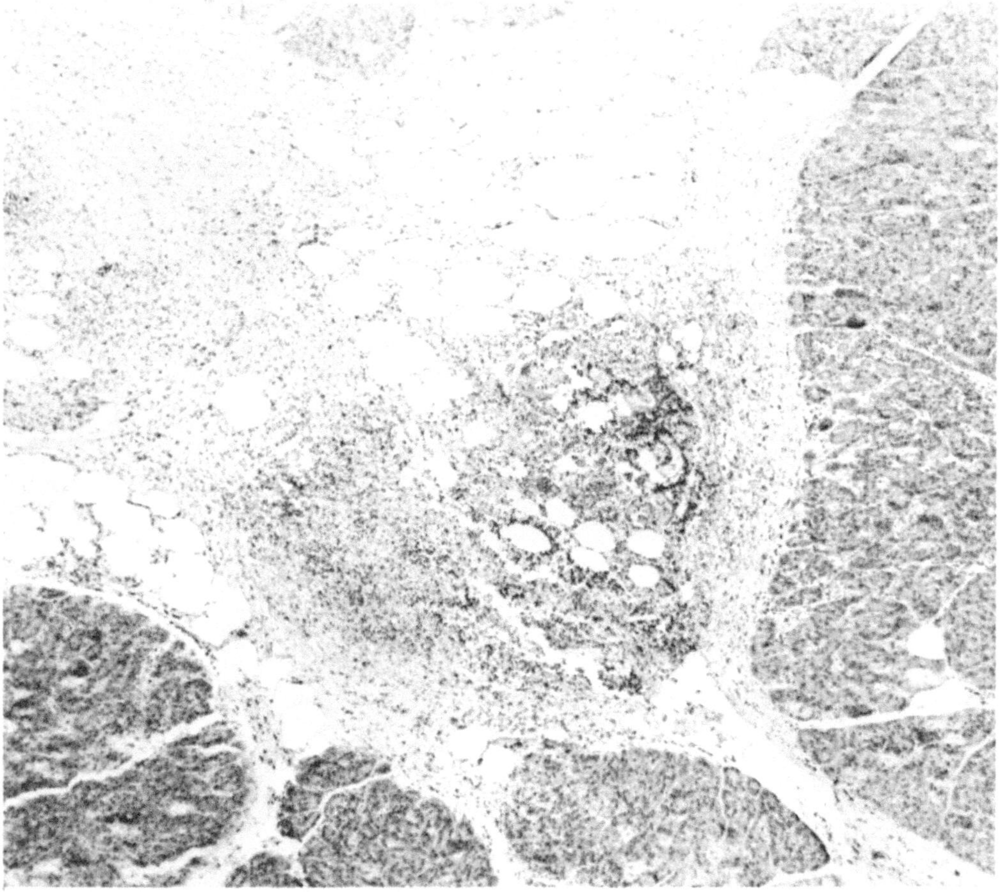

FIGURE 26. Histologic section of pancreas and adjacent fat, showing acute pancreatic necrosis of fat, in a sea lion with acute pancreatitis.

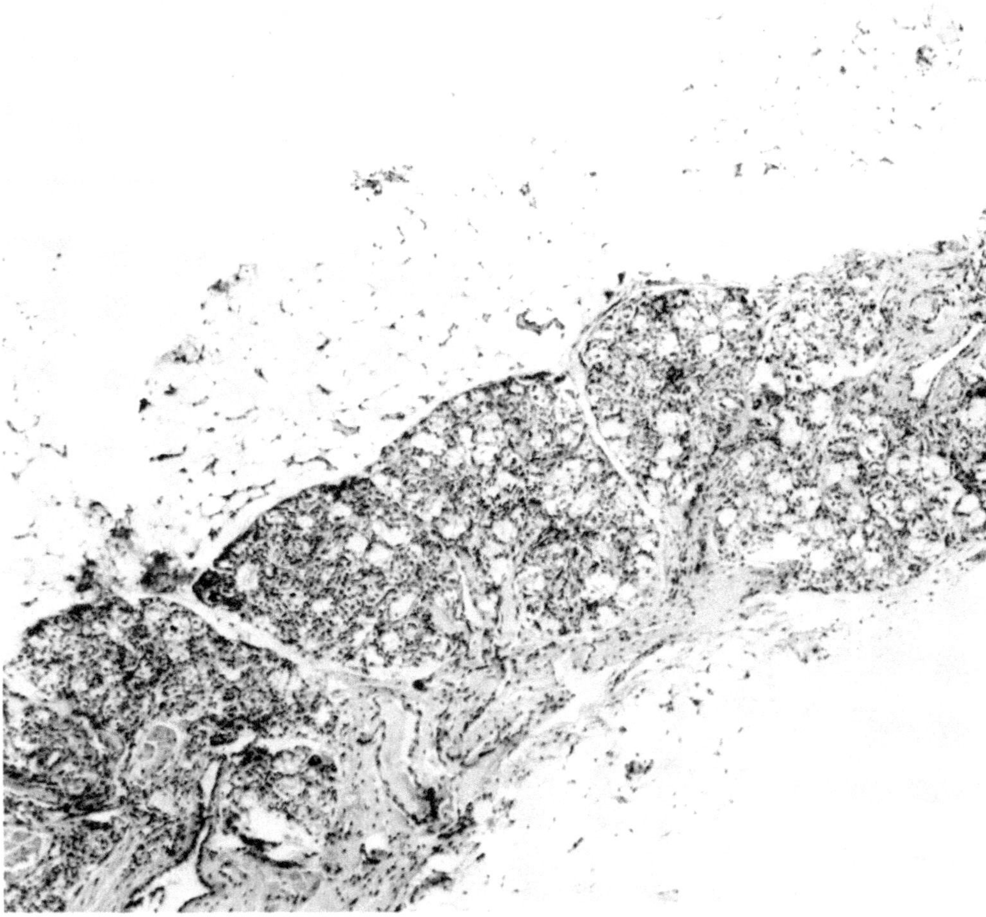

FIGURE 27. Section of pancreas from a sea lion. Grossly, the pancreatic tissue was a thin ribbon embedded in the fat. This was a case of severe pancreatic atrophy.

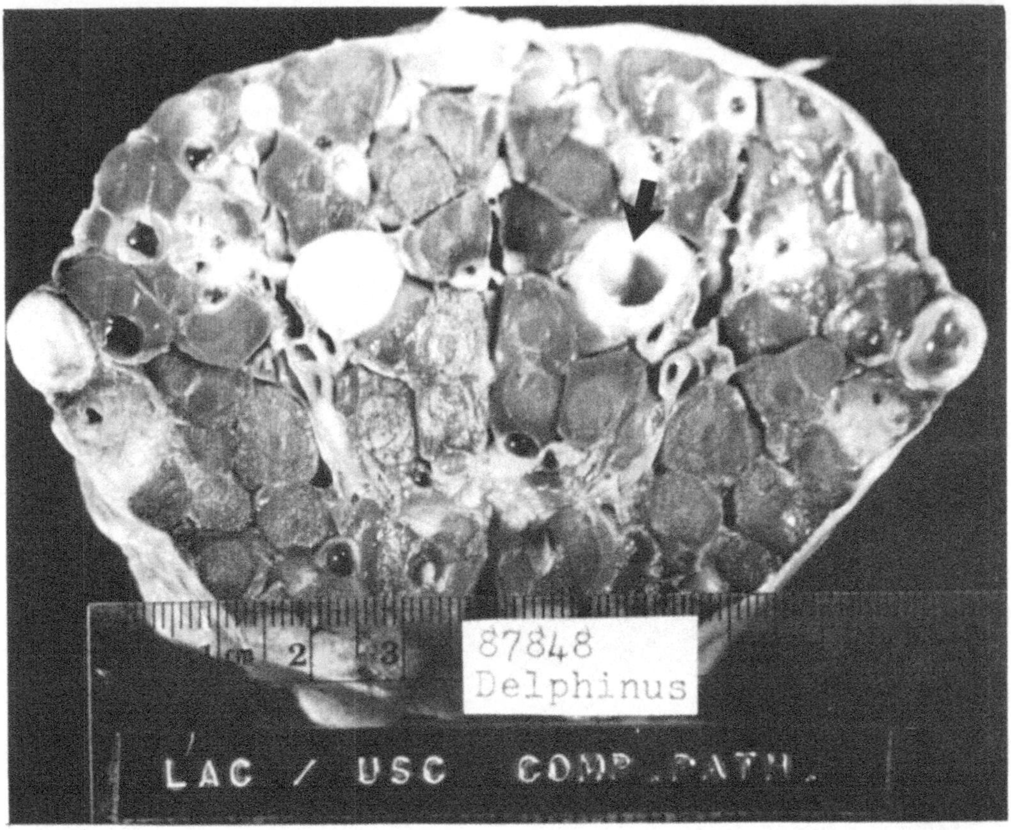

FIGURE 28. Gross cut section of the kidney from a common dolphin. There are numerous fluid filled cysts, and some of the cyst walls are very thick and fibrotic (arrows).

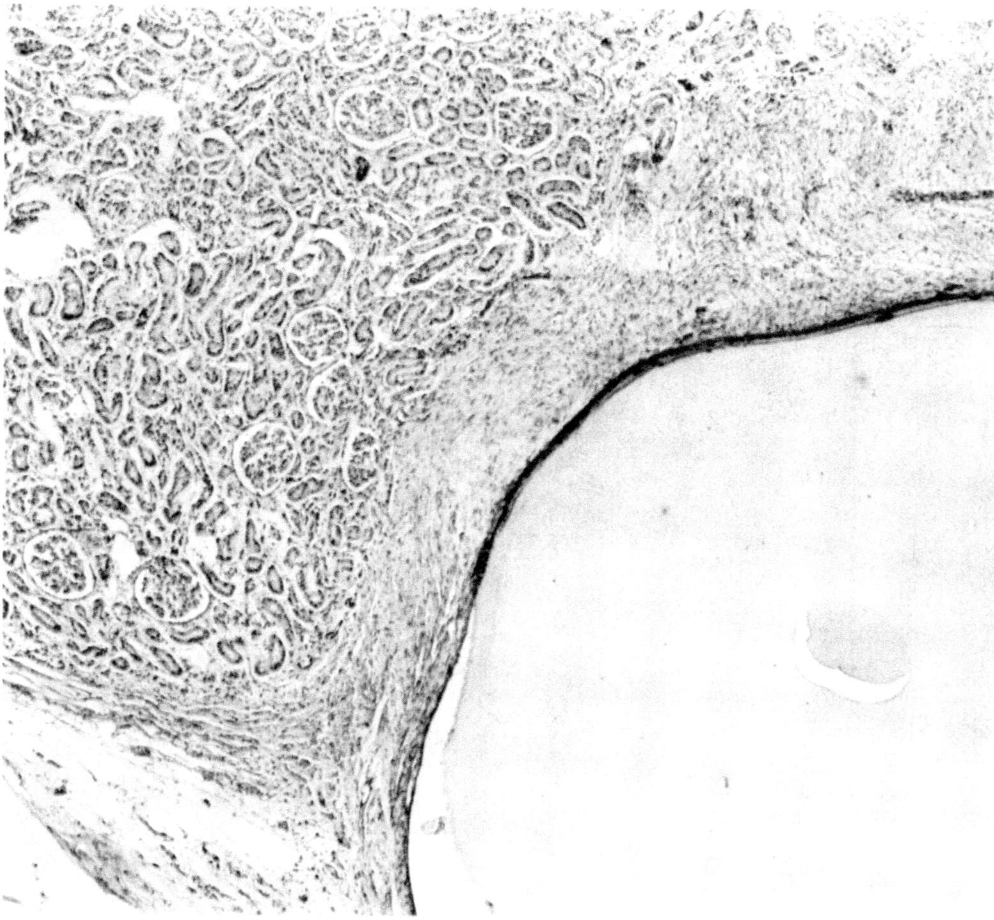

FIGURE 29. Photomicrograph of kidney section from a dolphin (Figure 28), showing thick walled fluid filled cyst compressing the adjacent renal cortex.

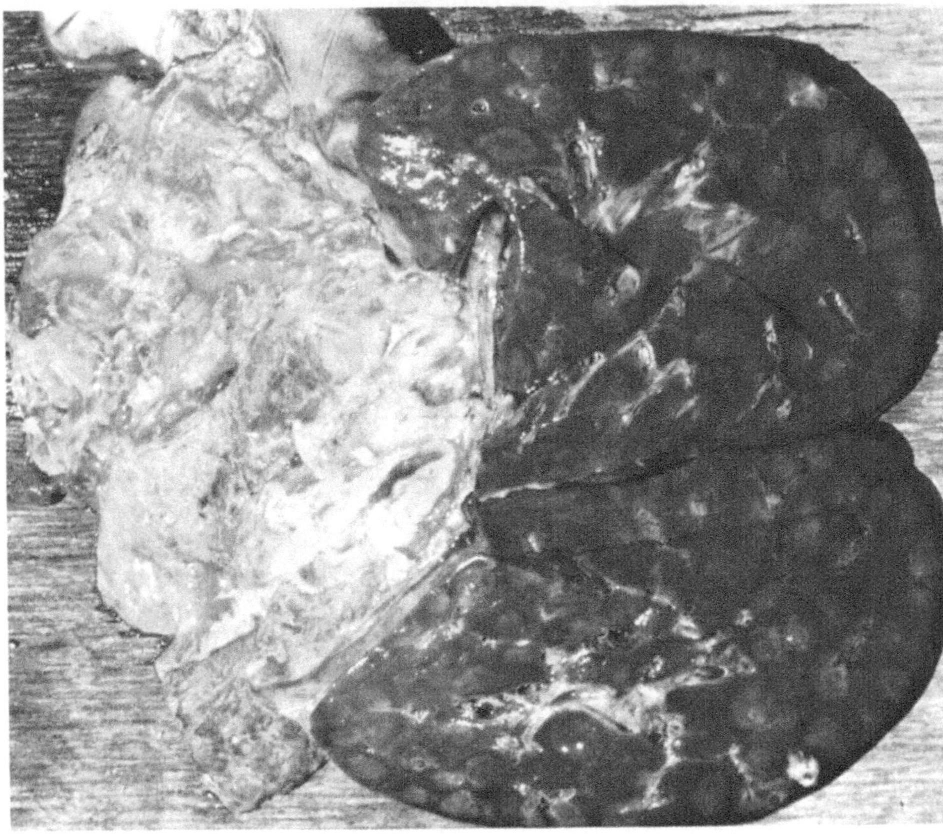

FIGURE 30. Kidney from a 5-year-old sea lion, with a large solitary cyst involving the anterior pole. The cyst has ruptured and collapsed.

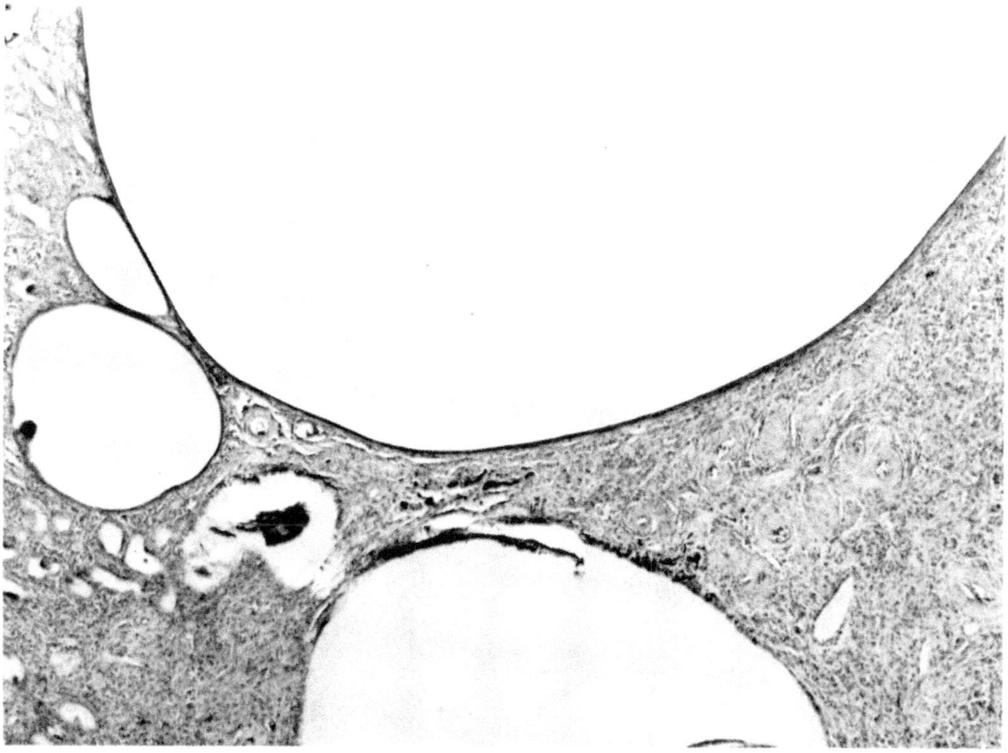

FIGURE 31. Congenital renal cysts in a sea lion. There are varying sized thin walled cysts in the cortex and medulla.

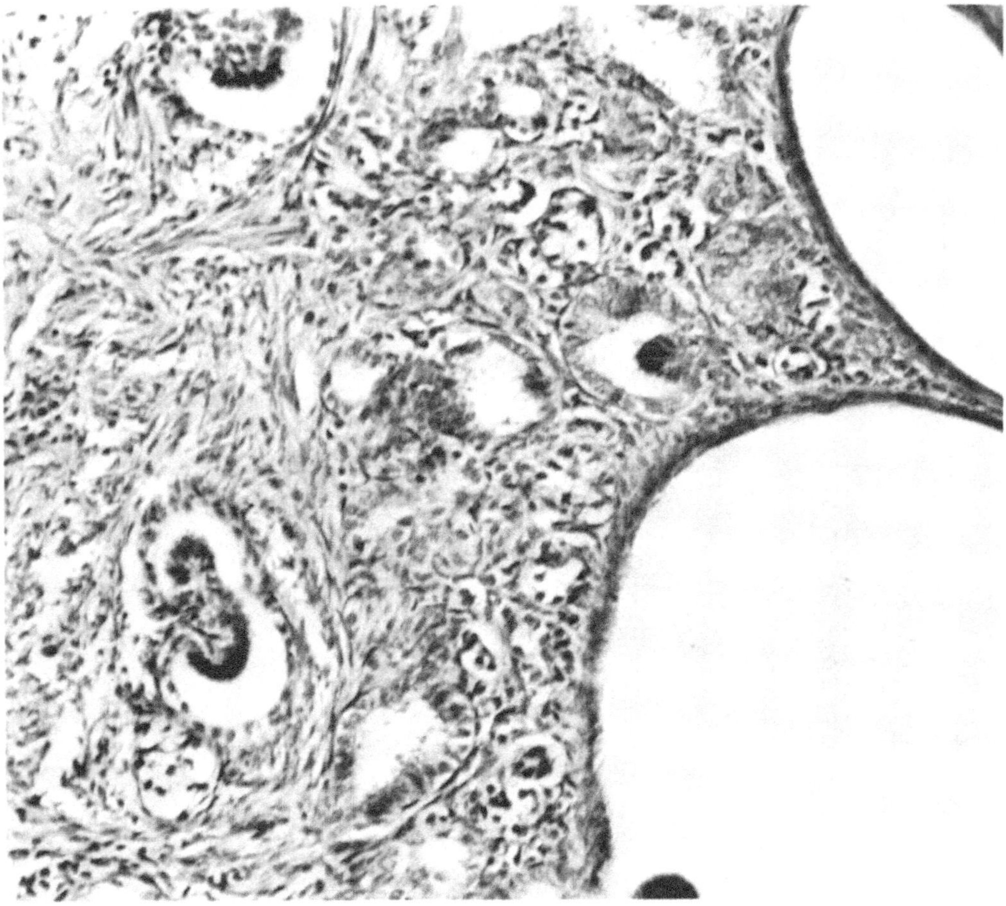

FIGURE 32. Higher magnification of kidney section seen in Figure 31, showing edge of cystic spaces, and immature glomerular tufts. This was believed to be a congenital anomalous condition.

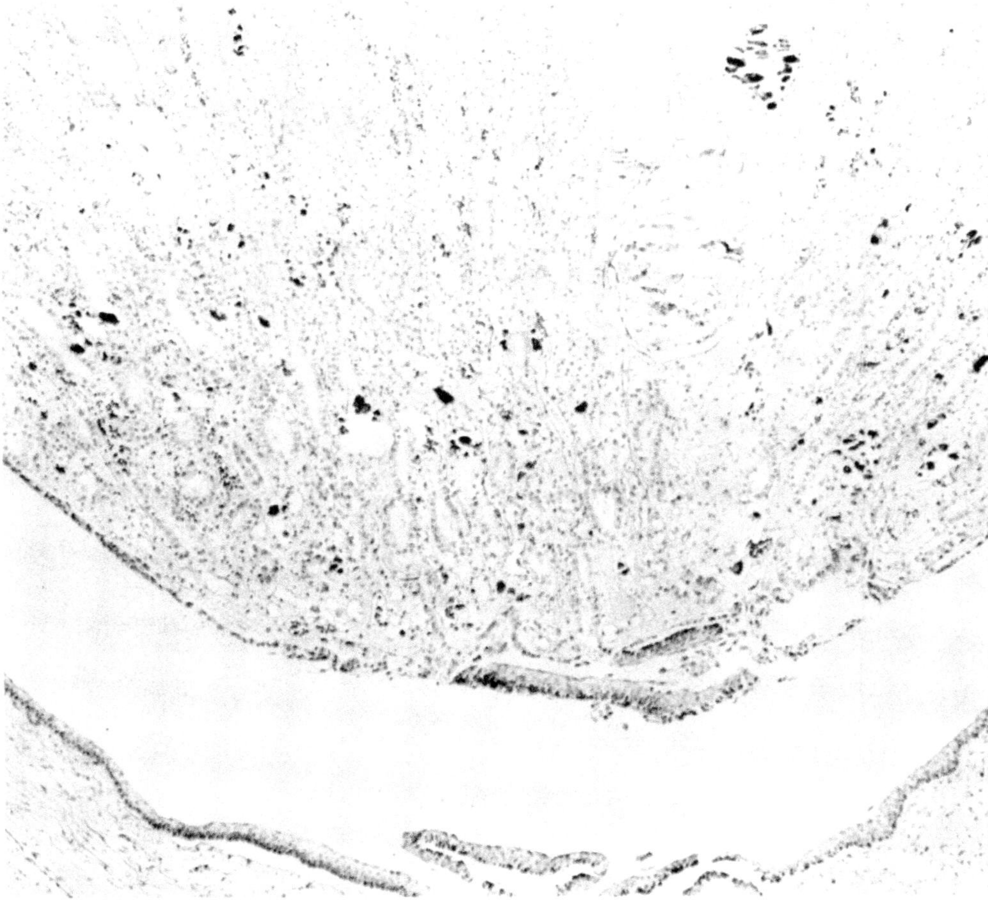

FIGURE 33. Renal papilla in a section of kidney from a beaked whale, showing numerous microcalculi within tubules. This appeared to be an incidental lesion.

FIGURE 34. Kidney from a harbor seal, showing multiple nephroliths in various calyces of the multiple reniculi of this pinniped kidney. (Courtesy of Dr. R. Stroud and the Comparative Pathology Registry, A.F.I.P.)

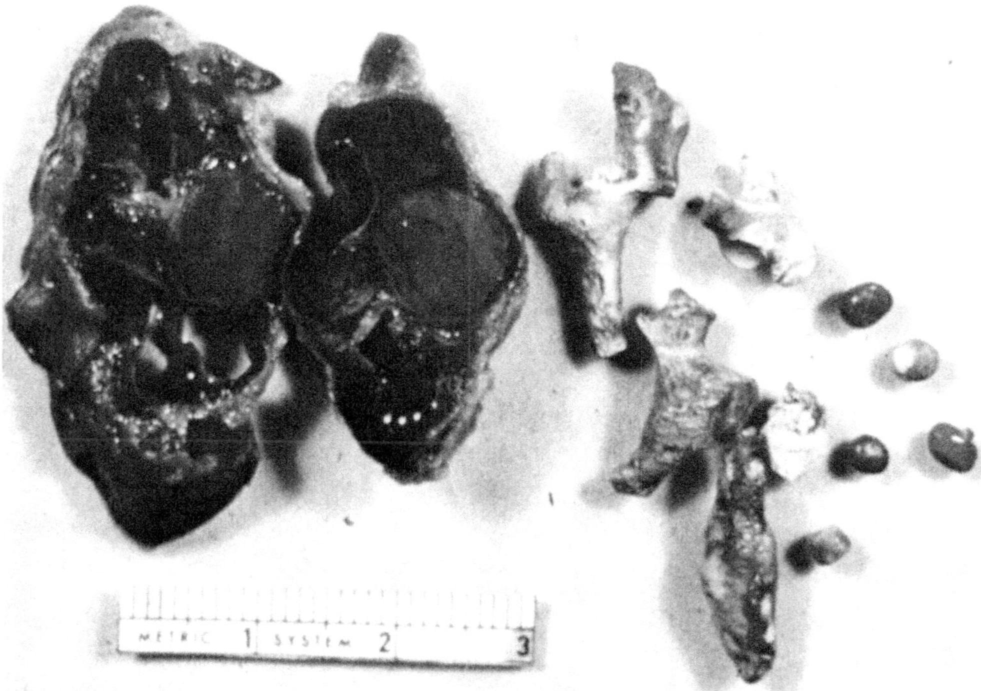

FIGURE 35. Multiple various shaped calculi removed from the kidney of a harbor seal (Figure 34). (Courtesy of Dr. R. Stroud and the Comparative Pathology Registry, A.F.I.P)

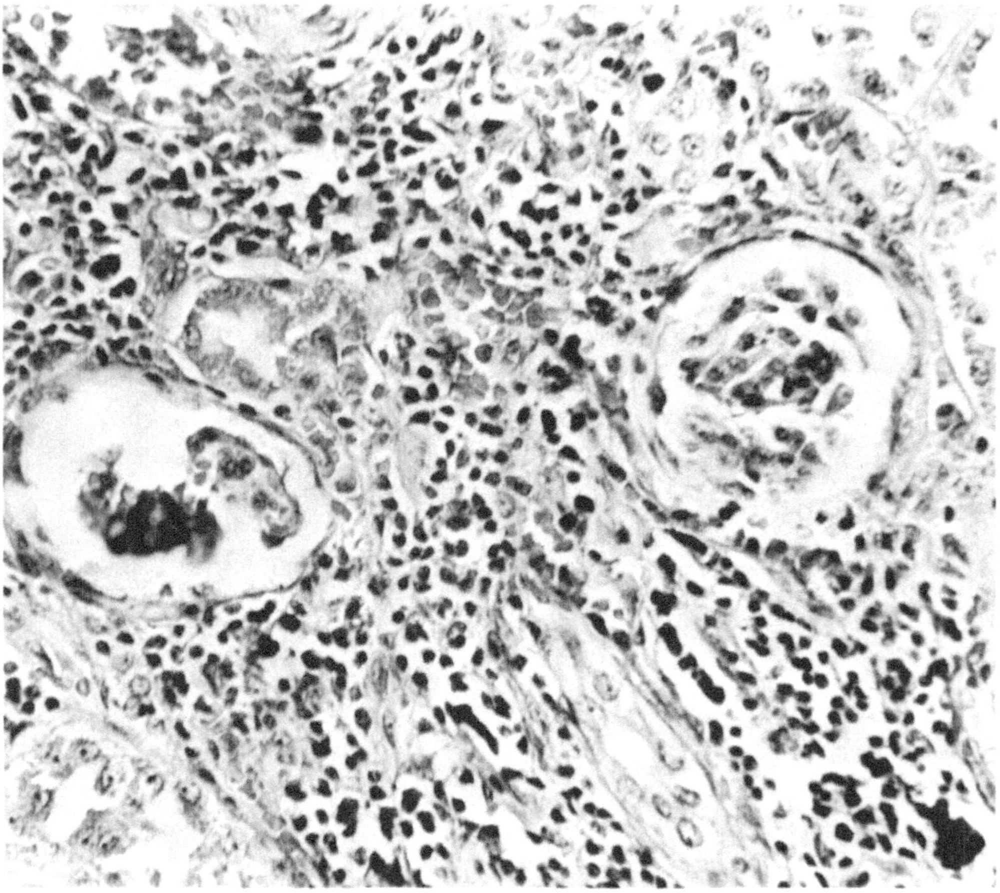

FIGURE 36. Section of kidney from a sea lion, showing the diffuse lymphoplasmacytic cortical infiltrate in a case of subacute interstitial nephritis.

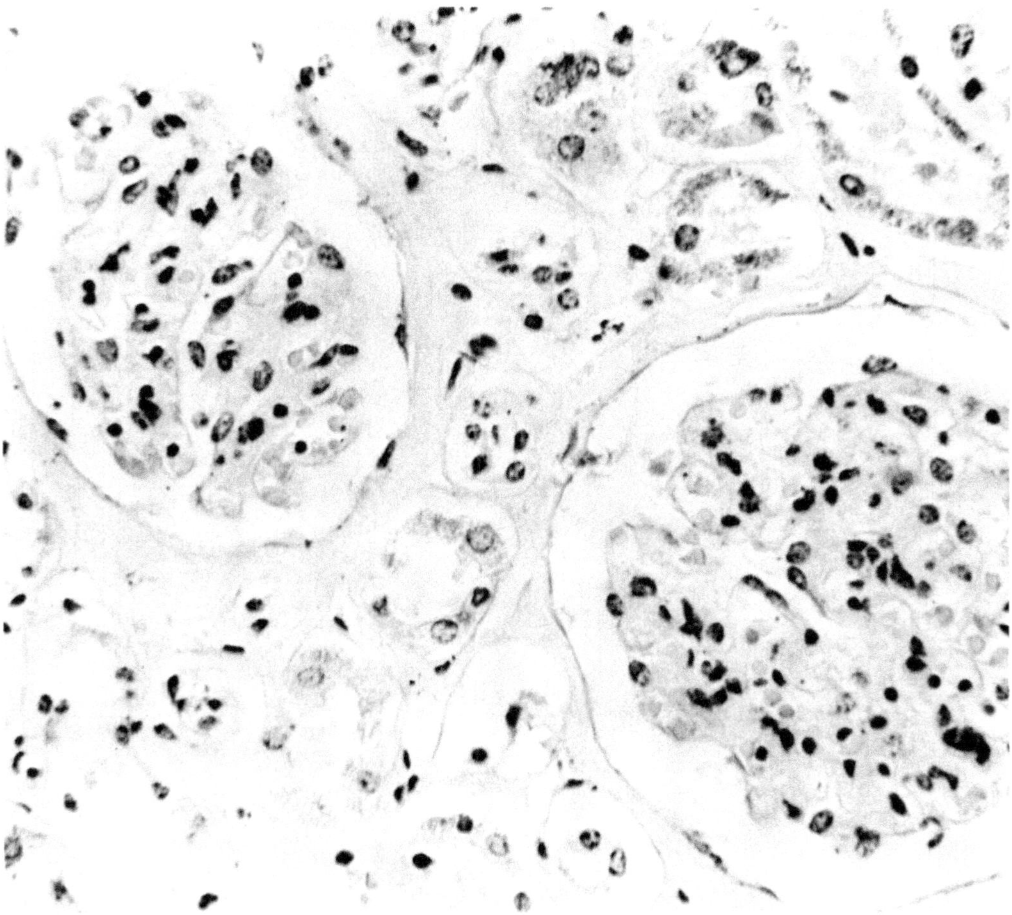

FIGURE 37. Kidney section from a beaked whale, demonstrating a moderate membranoproliferative glomerulonephritis. There is hypercellularity of the glomerular tufts, and a moderate thickening of Bowman's capsule.

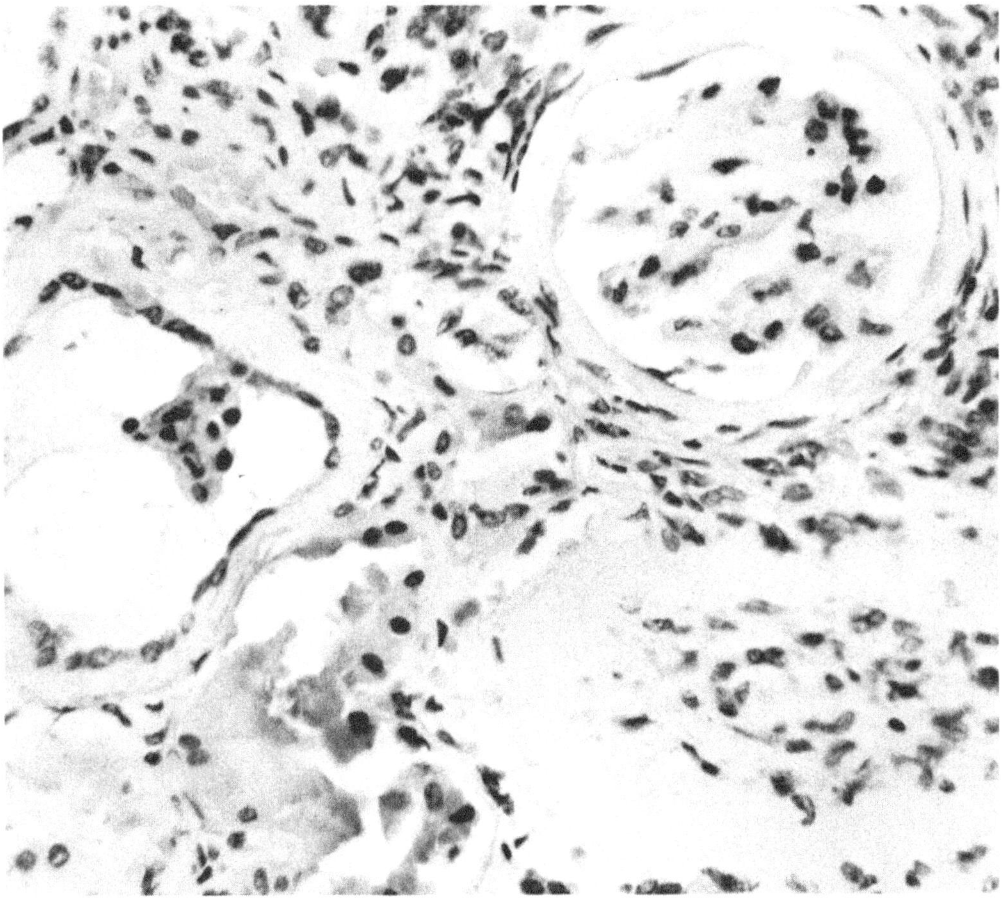

FIGURE 38. Section of kidney from a sea lion, with advanced glomerular disease. There is atrophy of glomerular tuft, hypertrophy of parietal epithelial cells, thickening of Bowman's capsule, and almost total sclerosis of one glomerulus.

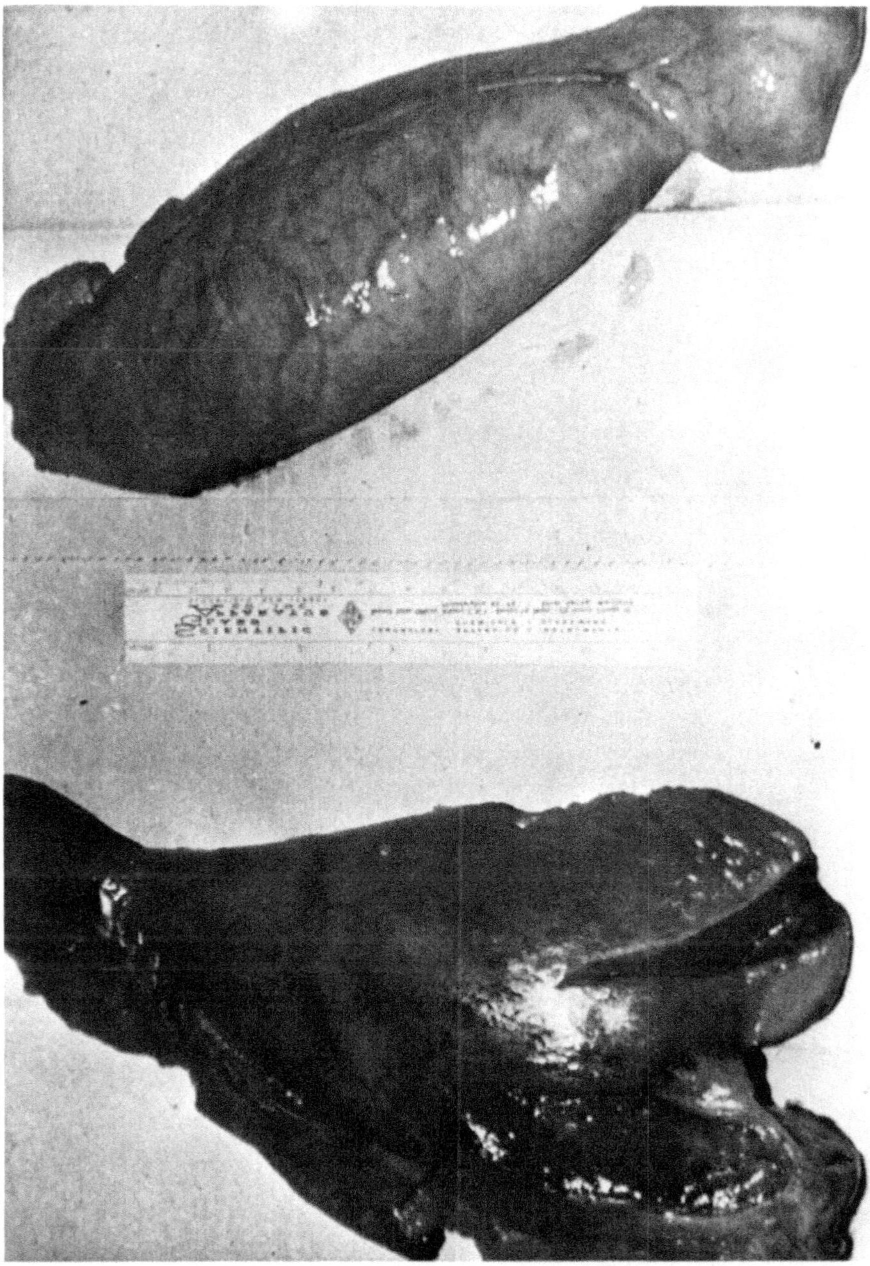

FIGURE 39. Testicles from a porpoise. The testicle on the left has acute orchitis with abscessation. It is markedly swollen, hyperemic, and contains an area of necrosis and softening.

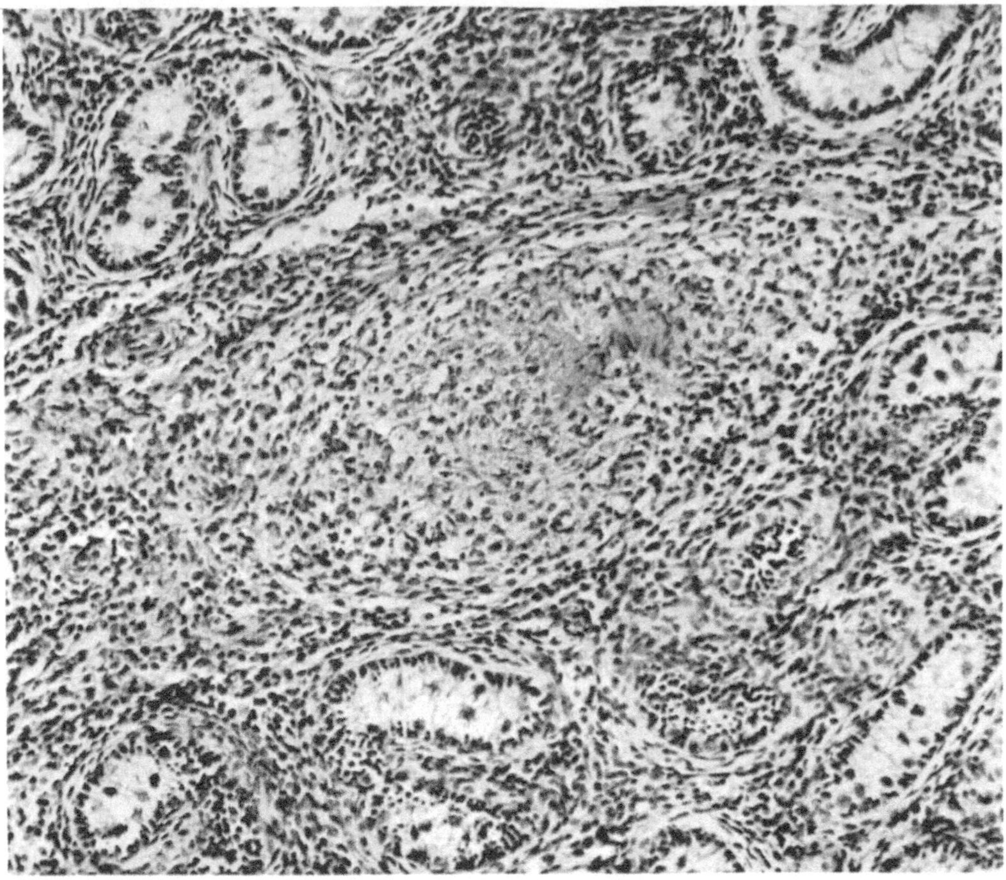

FIGURE 40. Granulomatous inflammation in the interstitium of the testicle of a porpoise. (Courtesy of Dr. J. Simpson and *Mammals of the Sea*, Charles C Thomas Publisher, 1972.)

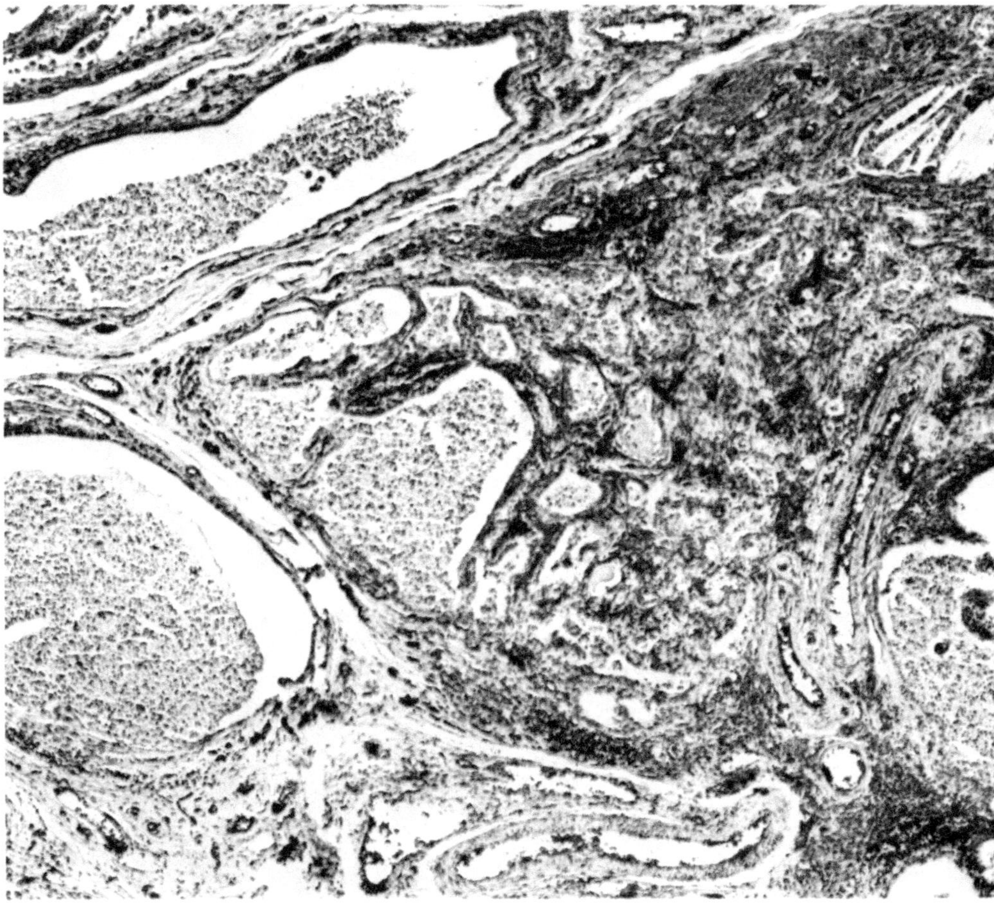

FIGURE 41. Thyroid gland from a northern right whale dolphin, showing marked cystic degeneration, debris filled cystic spaces, and inflammation.

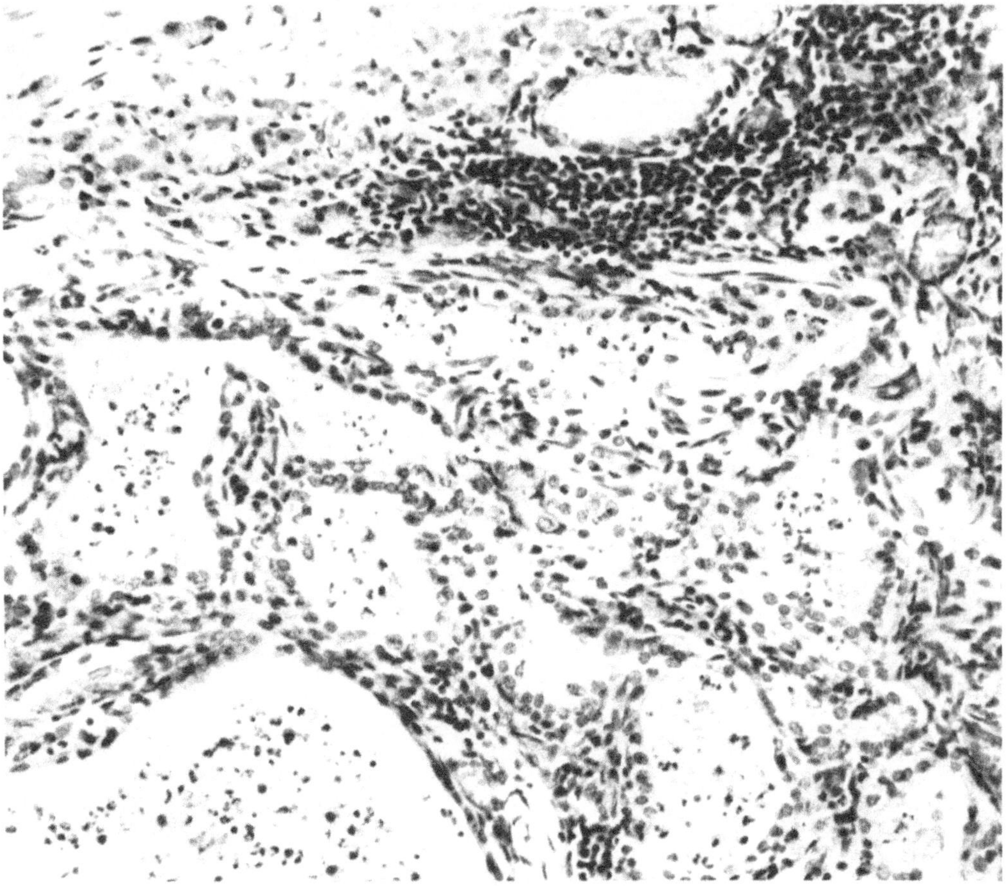

FIGURE 42. Higher magnification of thyroid gland in Figure 41, showing cystic degeneration of follicles, and a lymphocytic and monocytic infiltrate.

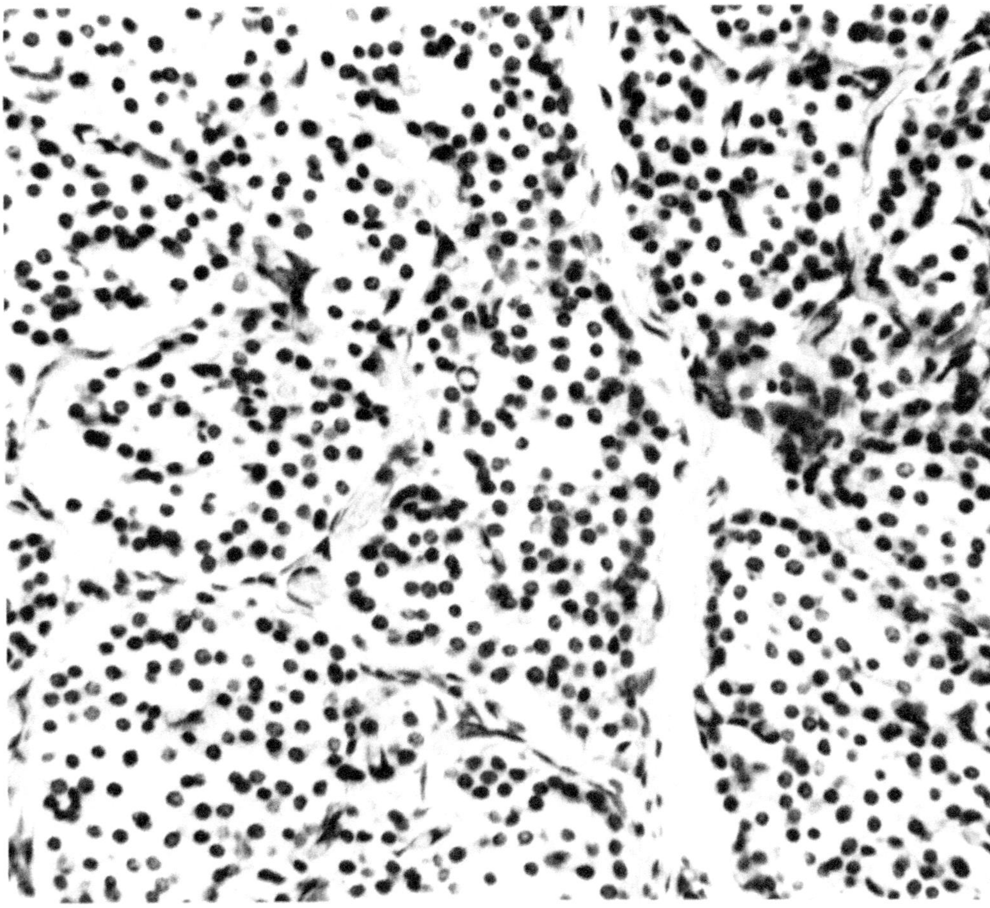

FIGURE 43. Thyroid gland from a dolphin, demonstrating modular hyperplastic proliferation of parafollicular cells. There is a distinct organoid pattern present.

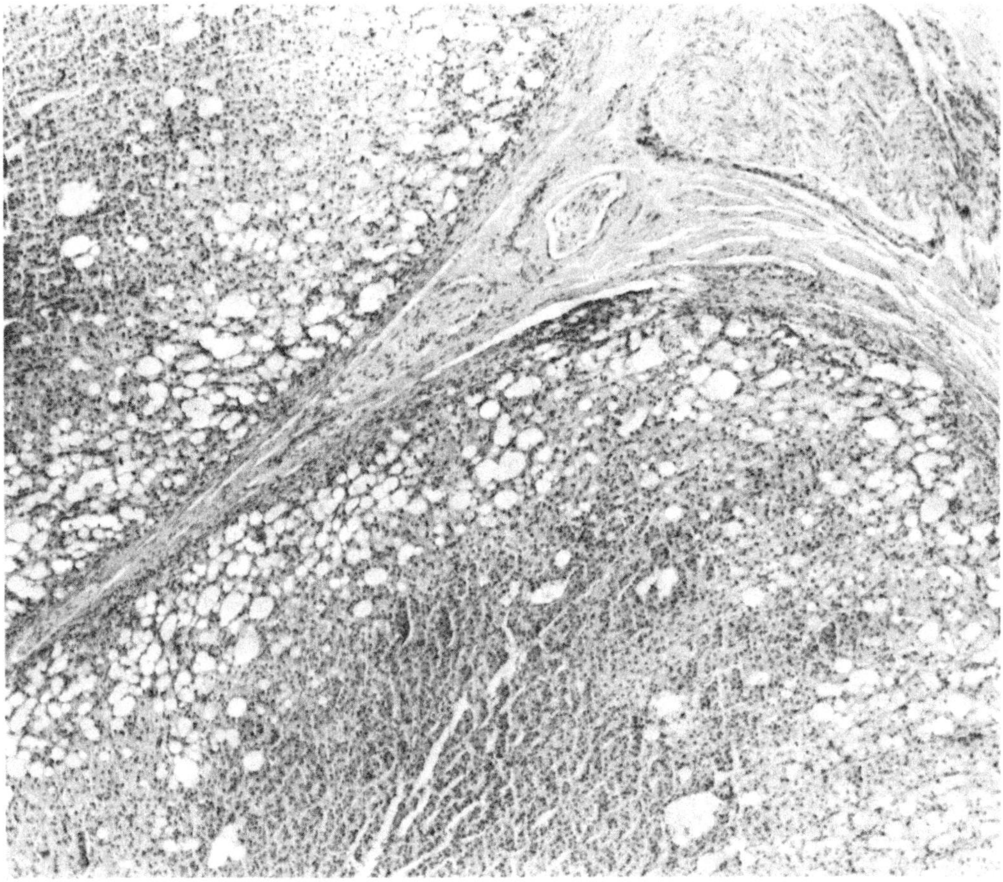

FIGURE 44. Vacuolar degeneration of the zona glomerulosa and outer part of the zona fasciculata. This may be a stress related lesion, often seen in adrenal glands of our series of animals.

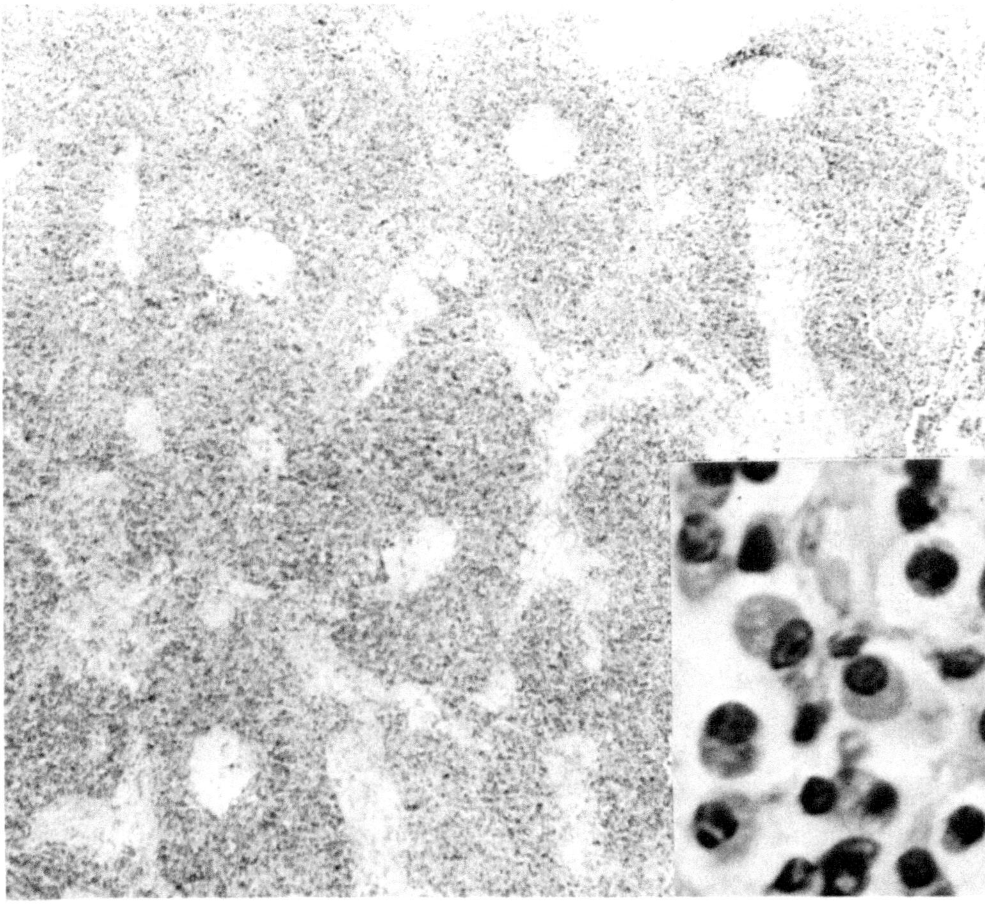

FIGURE 45. Spleen from a dolphin, showing amyloidosis of all detectable lymphoid centers. Inset shows plasma cell differentiation in the parafollicular areas.

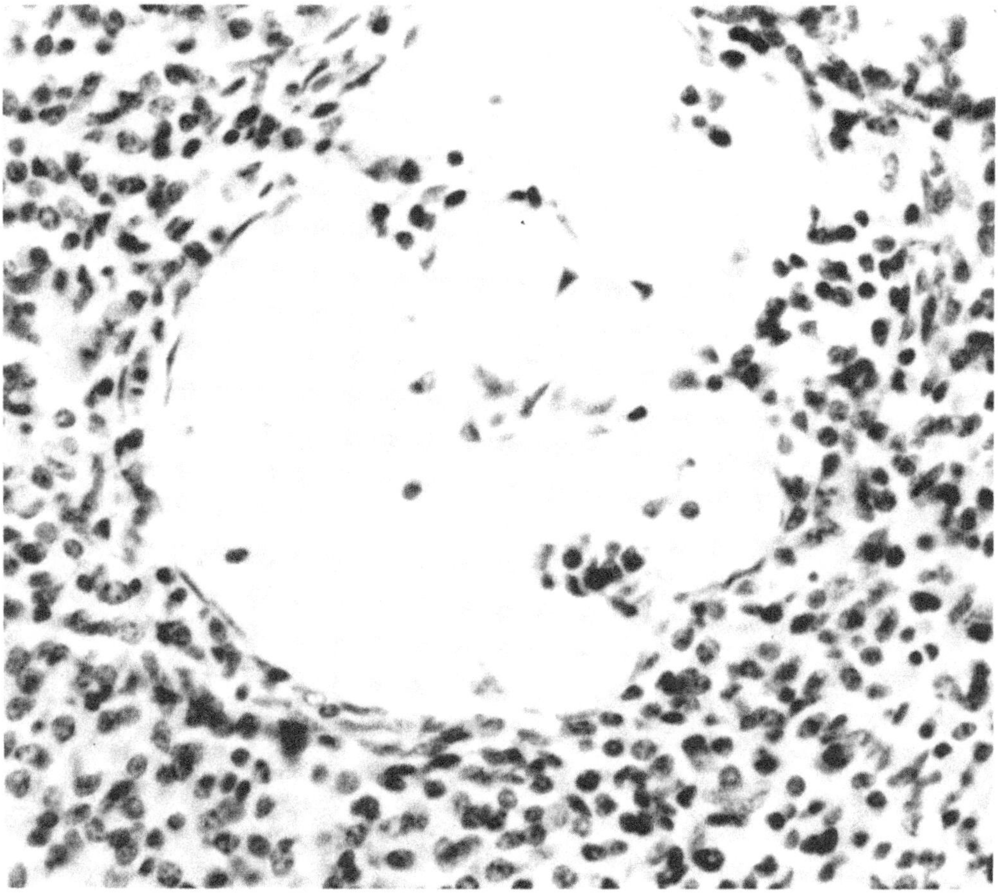

FIGURE 46. Higher magnification of lymphoid follicle of Figure 45, showing amorphous deposit of amyloid. This stained for amyloid by Congo Red and by birefringence.

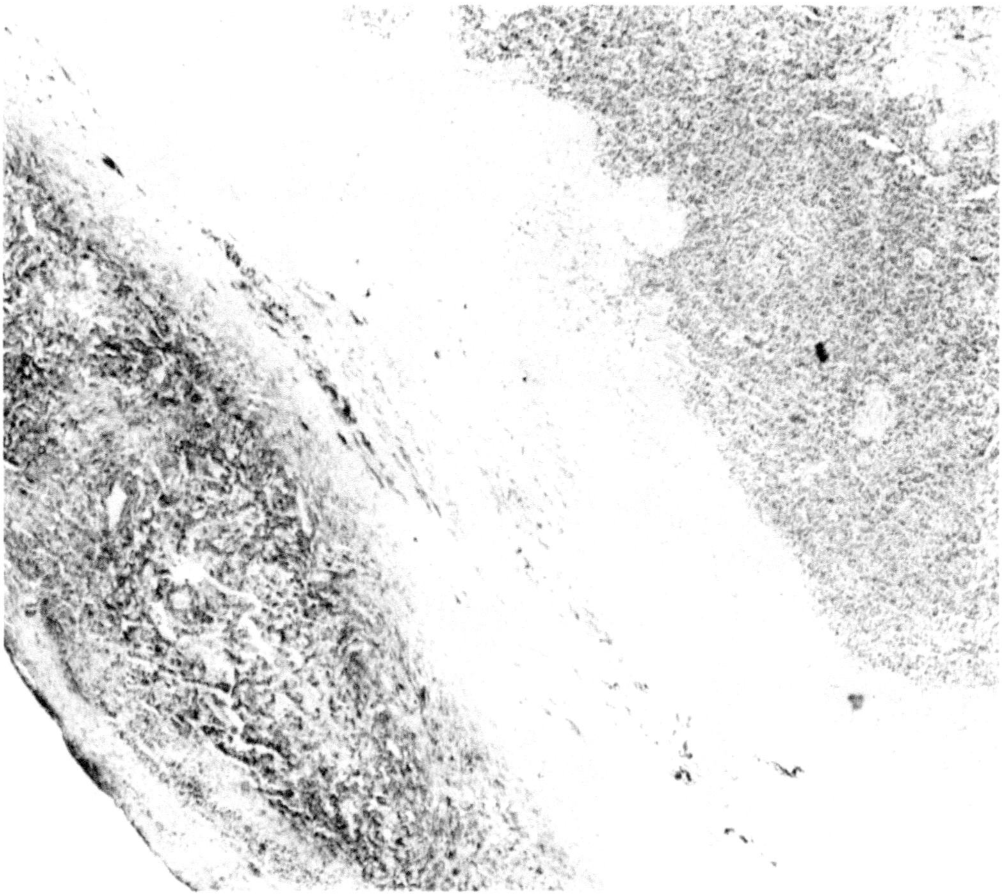

FIGURE 47. Capsule of spleen from a sea lion, demonstrating siderotic plaque and fibrosis, an area of prior hemorrhage (Gamna-Gandy body).

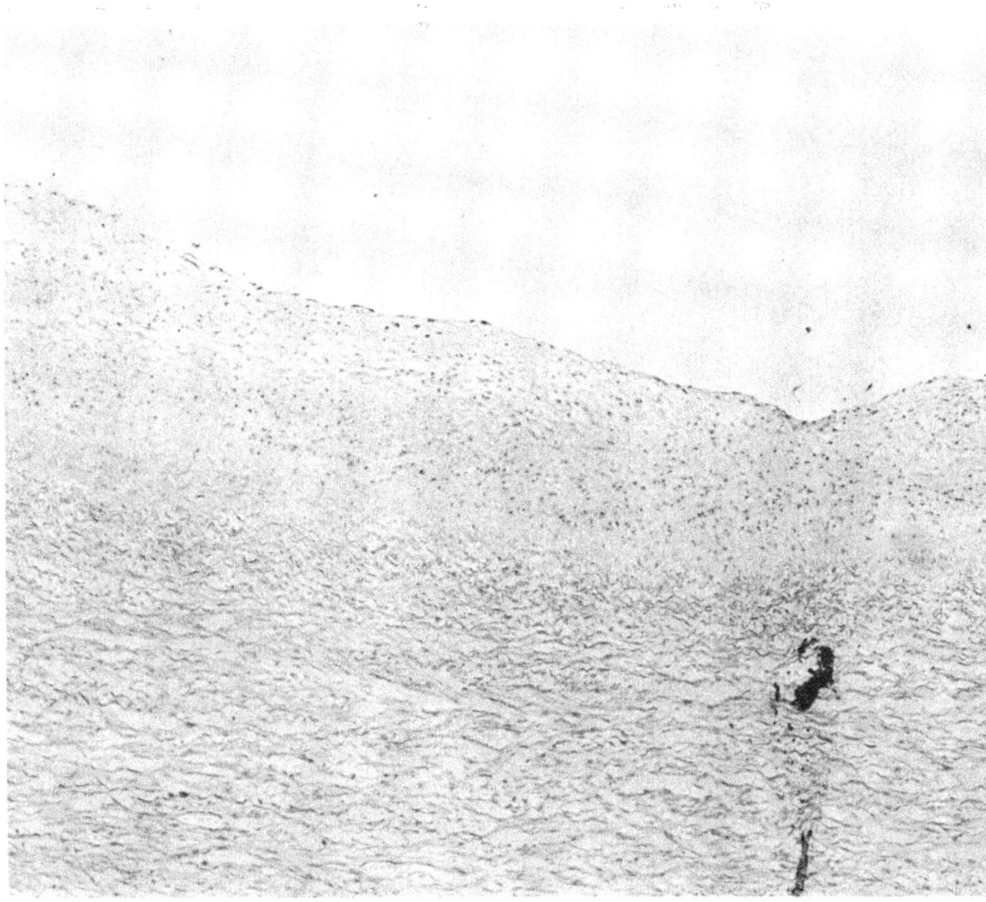

FIGURE 48. Section of aorta from a common dolphin, demonstrating a proliferative intimal plaque.

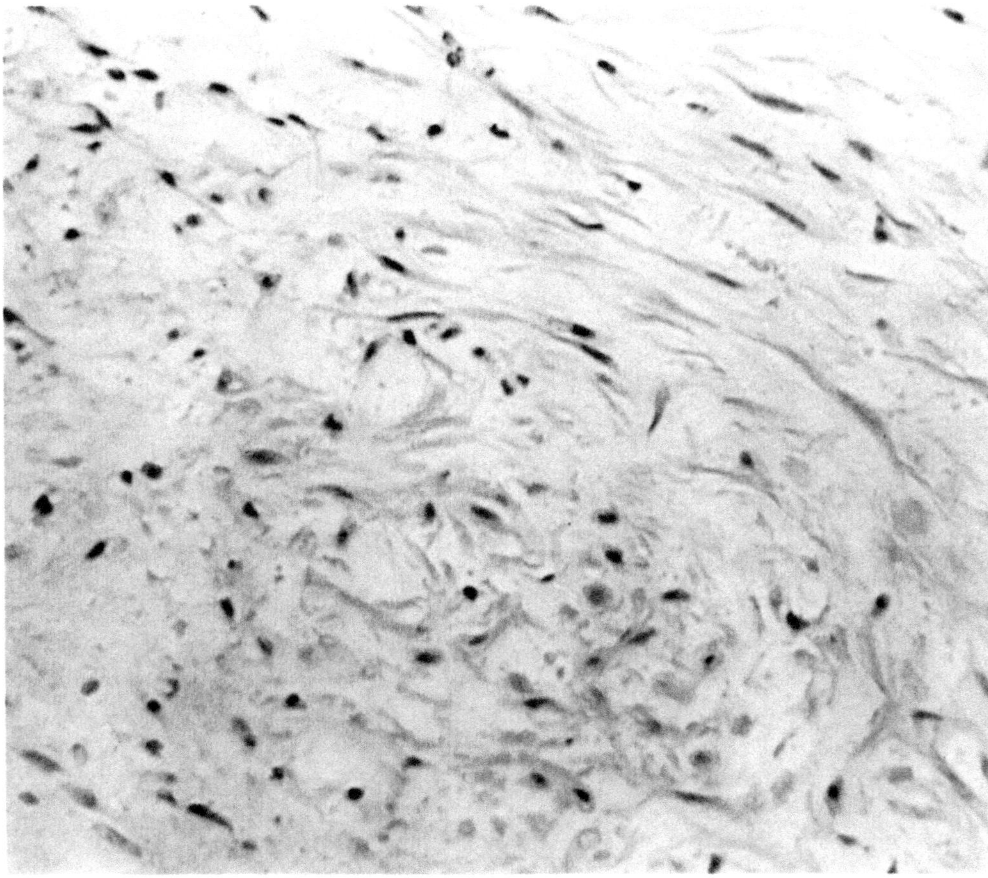

FIGURE 49. Higher magnification of area of plaque in Figure 48, showing collection of large lipid containing foam cells.

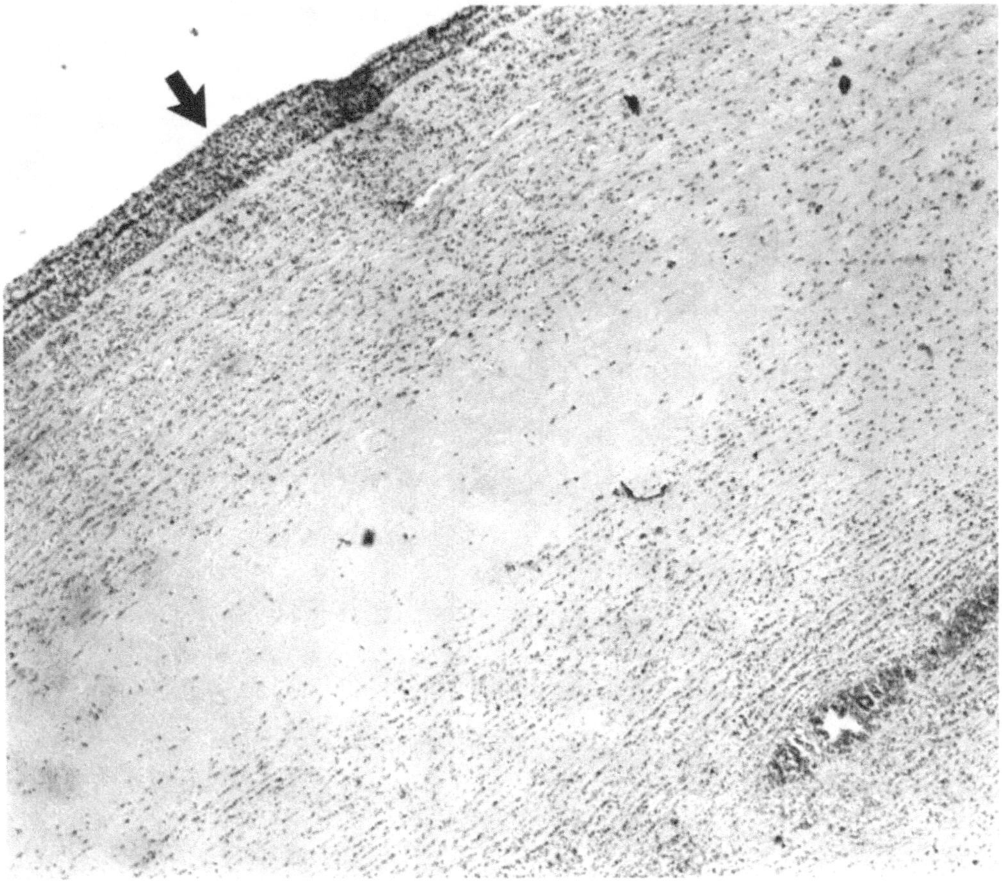

FIGURE 50. Intimal region of section of aorta from a Dall's porpoise, showing marked aortitis, with mixed cellular infiltrate (arrow).

FIGURE 51. Middle region of aortic wall from the Dall's porpoise of Figure 50, showing marked aortitis along the area of the vasa vasora.

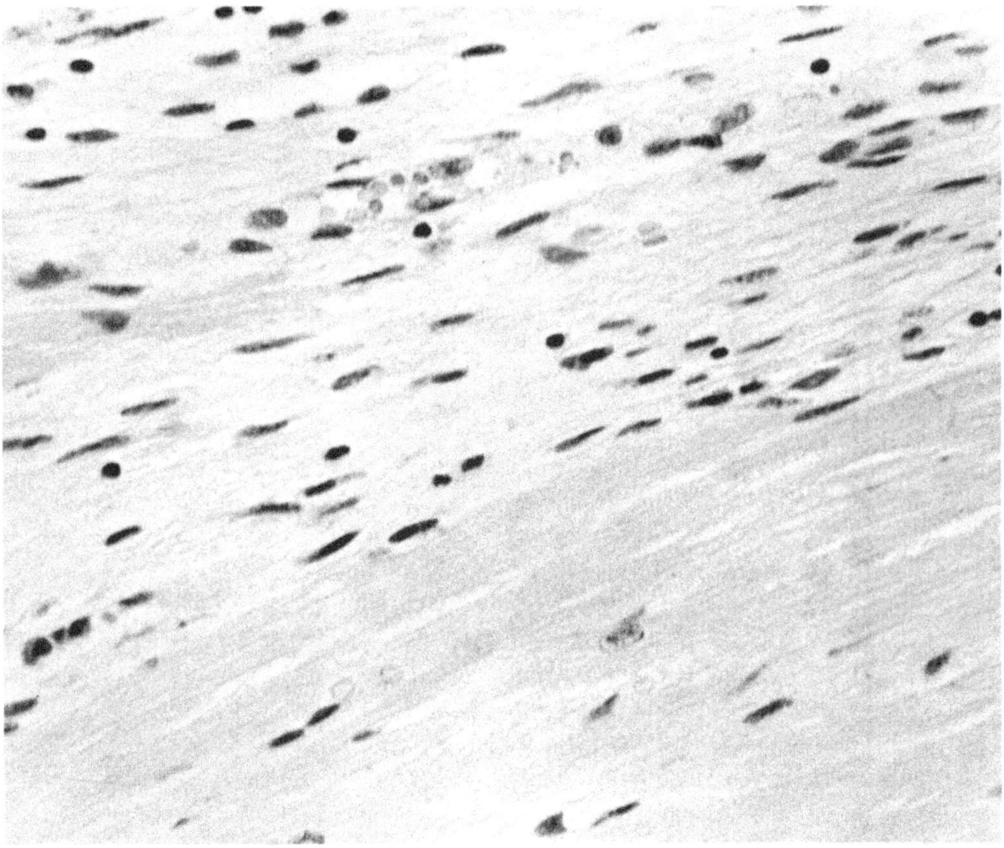

FIGURE 52. Section of myocardium from an adult female sea lion, demonstrating area of myocardial fiber loss and fibrosis.

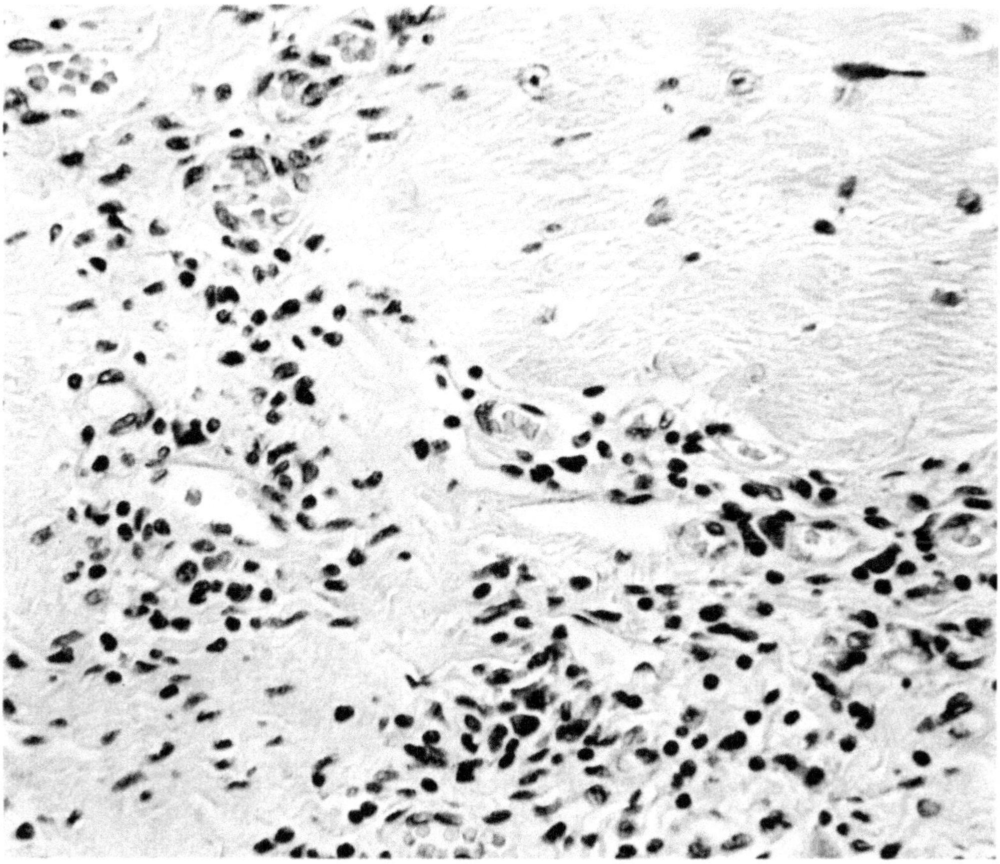

FIGURE 53. Another section of myocardium from the heart in Figure 52, showing active myocarditis, with mixture of mononuclear inflammatory cells.

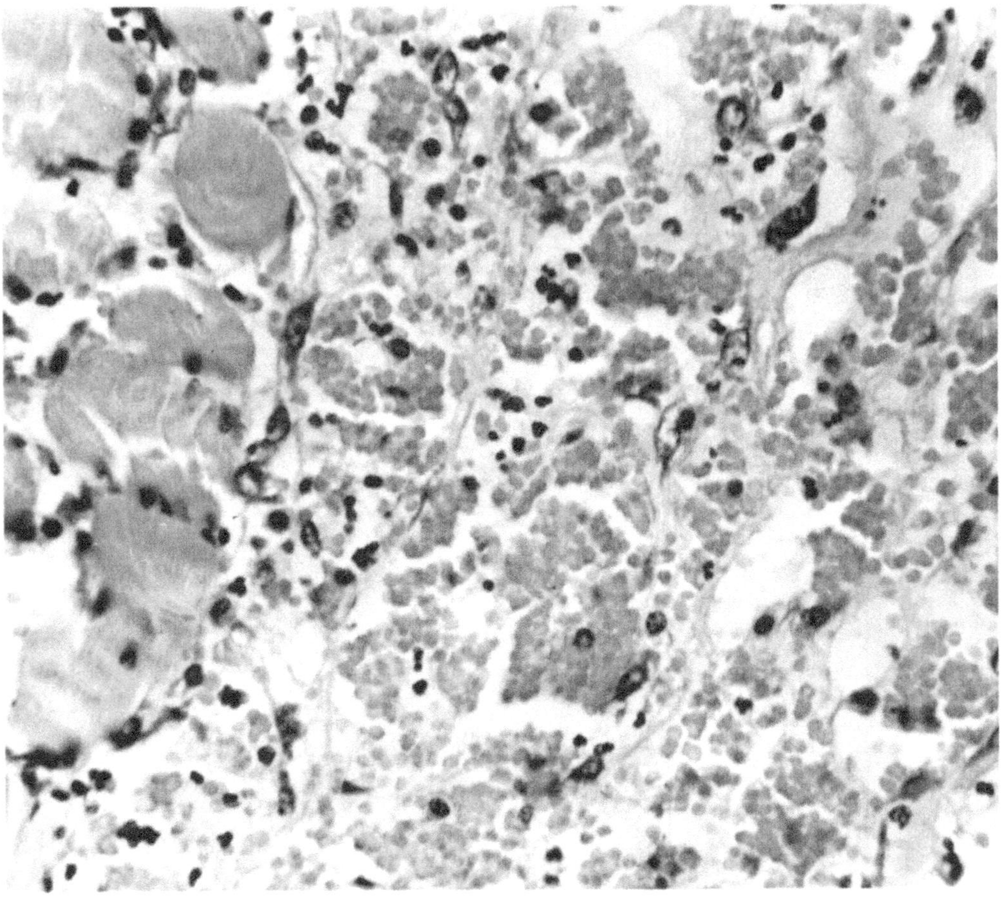

FIGURE 54. Section of skeletal muscle from a sea lion, demonstrating acute hemorrhage and necrosis of muscle bundles, in a lesion which is probably stress rhabdomyolysis (capture myopathy).

REFERENCES

1. **Simpson, J. G. and Gardner, M. B.,** Comparative microscopic anatomy of selected marine mammals, in *Mammals of the Sea, Biology and Medicine,* Ridgway, S. H., Ed., Charles C Thomas, Springfield, Ill., 1972, 298.
2. **Stroud, R. K. and Roffe, T. J.,** Causes of death in marine mammals, *J. Wildlife Dis.,* 15, 91, 1979.
3. **Ridgway, S. H. and Patton, G. S.,** Dolphin thyroid: some anatomical and physiological findings, *Z. Vergl. Physiologie,* 71, 129, 1971.
4. **Young, D. R., Tsu-Kai, J., Gossett, R. W., and Hershelman, G. P.,** Trace pollutants in surface runoff, in *Coastal Water Research Project Biennial Report for 1979—1980,* Bascom, W., Ed., Southern California Coastal Water Research Project, Long Beach, Calif., 1980, 163.
5. **Nishiwaki, M.,** General biology, in *Mammals of the Sea, Biology and Medicine,* Ridgway, S. H., Ed., Charles C Thomas, Springfield, Ill., 1972, 3.
6. **Ridgway, S. H., Geraci, J. R., and Medway, W.,** Diseases of pinnipeds, *Rapp. P.-V. Reun. Cons. Int. Explor. Mer.,* 169, 327, 1975.
7. **Sweeney, J. C. and Ridgway, S. H.,** Common diseases of small cetaceans, *J.A.V.M.A.,* 167, 533, 1975.
8. **Cowan, D. F.,** Pathology of the pilot whale, *Arch. Pathol.,* 82, 178, 1966.
9. **Roberts, J. C., Jr., Boice, R. C., Brownell, R. L., Jr., and Brown, D. H.,** Spontaneous atherosclerosis in Pacific toothed and baleen whales, in *Comparative Atherosclerosis,* Roberts, J. C., Jr. and Strauss, R., Eds., Harper and Row, New York, 1965, 151.
10. **Spraker, T. R.,** Pathophysiology associated with capture of wild animals, in *Comparative Pathology of Zoo Animals,* Montali, R. J. and Migaki, G., Eds., Smithsonian Institution Press, Washington, D.C., 1980, 403.
11. **Wilson, T. M.,** Diffuse muscular degeneration in captive harbor seals, *J.A.V.M.A.,* 161, 608, 1972.
12. **Leipold, H. W.,** Congenital defects of zoo and wild mammals: a review, in *Comparative Pathology of Zoo Animals,* Montali, R. J. and Migaki, G., Eds., Smithsonian Institution Press, Washington, D.C., 1980, 457.

INDEX

K

L

M

N

O

P

R

S

T

U

V

W

Z

For Product Safety Concerns and Information please contact our EU representative GPSR@taylorandfrancis.com Taylor & Francis Verlag GmbH, Kaufingerstraße 24, 80331 München, Germany

Batch number: 10731074

Printed by Printforce, the Netherlands